YOUR BODY
The Fish That Evolved

YOUR BODY
The Fish That Evolved

Dr Keith Harrison

Published by Metro Publishing
an imprint of John Blake Publishing Ltd,
3 Bramber Court, 2 Bramber Road,
London W14 9PB
England

www.blake.co.uk

First published in hardback in 2007

ISBN 978 1 84454 379 3

All rights reserved. No part of this publication may be reproduced,
stored in a retrieval system, or in any form or by any means, without the
prior permission in writing of the publisher, nor be otherwise circulated
in any form of binding or cover other than that in which it is
published and without a similar condition including this condition
being imposed on the subsequent publisher.

British Library Cataloguing-in-Publication Data:

A catalogue record for this book is available from the British Library.

Design by www.envydesign.co.uk

Printed in Great Britain by Creative, Print & Design

1 3 5 7 9 10 8 6 4 2

© Text copyright Dr Keith Harrison

Papers used by John Blake Publishing are natural, recyclable products
made from wood grown in sustainable forests. The manufacturing processes
conform to the environmental regulations of the country of origin.

Every attempt has been made to contact the relevant
copyright-holders, but some were unobtainable. We would be
grateful if the appropriate people could contact us.

Preface

There is no aspect of nature more obvious or more personal than our own body, but how much do we really know about it? Why do we have two arms and two legs, not four arms or six legs? Why do we have ribs across our chest but not across our stomach? Why do our elbows and our knees bend in opposite directions (and have you ever noticed)? This book sets out to answer these questions and others as it traces the evolution of every one of us, not from our cousins the apes but from our more distant ancestors – the fish.

Contents

1	The Human Pedigree	1
2	Science, Religion and Rocks	7
3	Evolution, Darwin and Natural Selection	15
4	Genes	25
5	Evolution in Practice	37
6	When We Were Fish	49
7	When We Were Amphibians	63
8	When We Were Reptiles	69
9	Mammals	77
10	Primates	97
11	'Hominids'	111

12	**Your Body Today**	117
13	**Your Body's Problems**	153
14	**Your Brain**	167
15	**Future Evolution of the Human Body**	193

Chapter One

The human pedigree

The story of our bodies did not begin when our ape-like ancestors left the trees. By then, it was already a long tale stretching back before the evolution of the first fish, 500 million years ago. We are descended from those fish, as is every other animal with a backbone that has ever lived, from the smallest frogs and lizards to the largest elephants and dinosaurs.

Once fish appeared in the ancient seas, they became increasingly widespread. Some moved into fresh water, then onto the land. Natural selection did its work and the first amphibians evolved. Some amphibians turned into the first dry-land vertebrates, the reptiles, and, while one group of reptiles evolved larger and larger bodies and became the dinosaurs, others became the first mammals and grew smaller and smaller. When the dinosaurs faded, leaving only their descendants the birds to dominate the air, mammals took over the ground and the trees. Eventually, one group of mammals stood upright and walked out of the forest. The rest, as they say, is history.

This is the story of what created that history. We will explore our time as fish and will track evolution through our amphibian and reptile pasts to our life as mammals. Each stage of this journey has left its mark on our bodies and to understand why we look the way we do today we must first understand where we came from.

As vertebrates, we can trace many important parts of our design back to the first fish, but our overall shape is even older than that.

Five hundred million years ago, the seas were teeming with animals but every one was a type of invertebrate. Today, we are familiar with many of their relatives: insects, arachnids and crustaceans (with bodies encased in hard jointed shells); molluscs (including clams with two hinged shells, snails with a spiral shell, garden slugs and squid with an internal shell, octopus with no shells); echinoderms, well named for their 'prickly skin' (starfish, sea-urchins, sea-cucumbers); segmented

worms and their relatives (earthworms, rag-worms, lugworms, leeches); unsegmented roundworms and flatworms; sea-anemones, corals, jelly-fish; and other groups less well known and too numerous to list.

In these ancient seas filled with invertebrates, an innovation appeared that was to change the face of nature forever. A species evolved a stiffening rod down the centre of its body. It had become a fish. This rod would later be turned by natural selection into a row of bones, the vertebrae, and we vertebrates had started our epic journey. Scientists still don't know which invertebrate group we have to thank for our backbone, a structure so dominant in our bodies and in our minds that we refer to it in the singular even though it has more than 26 separate bones, and which we cite as the epitome of strength: 'Show some backbone. Are you spineless?' However, we can say something about our invertebrate ancestor's body.

What we inherit from our invertebrate ancestors

Animal bodies can take various forms. Some radiate outwards in all directions from the centre like a starfish or a coral polyp, but most have sides that are mirror images of each other. Whatever they have on one side they also have on the other side, and many of the organs occur as pairs. Those parts of the anatomy there is only one of, like the intestine, usually lie along the mid-line.

The invertebrate that became a fish was one of these bilaterally symmetrical forms. Every vertebrate that has ever lived has consequently followed the same pattern, including us. We have paired arms and legs, eyes, ears, nostrils, lungs, kidneys, ovaries and testes (testicles), and in the centre-line of our bodies we have one brain (with some paired aspects), one spine, one heart (which leans to the left), one reproductive organ and one intestine (greatly coiled so it can be up to six times longer than we are) with its one entry and one exit.

Our early invertebrate ancestors were apparently animals that moved through their environment as we also inherited from them a head and a tail, although since we stood upright these have become the top and the bottom. Any animal that moves – whether it's a worm, a lobster or a snail – has evolved sense organs at its front end, the end that encounters the environment first. Having all your sense organs on your tail would not have much survival value. An animal needs to know that it's about to crawl into a predator's mouth not that it's just finished crawling into a predator's mouth. For a similar reason an animal's mouth is usually at the front of the body so it's the first thing to encounter food. This is especially

important for predators where the food may be able to escape if it gets some warning (lions wouldn't get many dinners if they backed into zebras tail-first).

This sensible arrangement has led in almost all animal groups to the evolution of a head, which we wear as a curiously shaped stalked ball balanced on the top of our torso, but which other animals have kept at the front. Most of our sense organs are there: sight, smell, taste and hearing, and that's where we put our food. With so much information passing to the nerves from these sense organs, the processing of this information also takes place in the head. That's why the brain evolved there. We owe all these fundamental aspects of our body to our invertebrate past.

Timing

This book has hardly begun and already I'm talking casually about evolution without having said anything about it. Before we continue the story and describe our time as fish, let's pause to explore in simple terms the timescales involved and the ideas of science, evolution and natural selection that underpin our understanding of ourselves. We can begin by putting the evolution of life in perspective.

The Earth is about 4,550,000,000 years old. If we compress the whole of this time into one year, with the Earth forming on 1 January and today being midnight on 31 December, the first microscopic living cells appeared on 1 March, but the ancestral fish – those earliest vertebrates – did not appear until 21 November. It had taken about 750 million years for life to evolve from simpler chemistry, then more than another 3,000 million years (two-thirds of the age of the Earth) for that life to create the complexity of the fish. After that, things changed quickly but it was December before some fish colonised the land. Amphibians appeared on 2 December, followed by reptiles on the 8th. Mammals appeared on the 13th and the dinosaurs died out just after tea on the 26th. Humans did not arrive until this evening, just a few hours ago.

Chapter Two

Science, Religion and Rocks

In this book, we are going to explore the history of the human body. As everything we shall meet has been discovered by generations of scientists, it is worth taking a minute or two to consider what exactly this thing science is.

The word 'science' is just Latin for 'knowledge' but throughout history the way people have decided what they know about the universe hasn't always been constant. In medieval Europe, scholars would observe the world around them and theorise about why it was the way it was. They would then gather and debate their theories in an attempt to persuade others of their viewpoint. This convention of arguing the way to an agreed explanation eventually went out of fashion to be replaced in the 17th century by the rise of the scientific method.

The scientific method is an approach to understanding which we can visualise as a triangle. First, we observe the universe (or more usually the part that interests us). Next, we devise a theory to explain what we are seeing – a hypothesis. So far, this does not differ from the old approach but now a new step was added. Instead of arguing about the strengths and weaknesses of the theory, we test it in some way, usually by carrying out a form of experiment. When we view the result of the experiment, we are back at the beginning of the triangle, again making an observation.

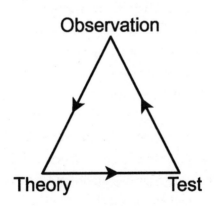

The Scientific Method

We can go round this triangle as many times at it takes to convince ourselves that we finally understand

SCIENCE, RELIGION AND ROCKS

what is happening, modifying the theory and devising new tests each time.

The scientific method now predominates in most cultures but it was not a new invention. It was simply an extension of how we live our everyday lives. For example, imagine you are walking down the street and see a fist-sized fuzzy brown ball on the path ahead of you. This is an observation (Step 1). You wonder what it is and speculate it might be a small coconut from the nearby market. You now have a theory: 'It's a coconut' (Step 2). You bend down to examine the ball and you roll it over with your foot. You are now conducting an experiment to test the theory (Step 3). As you observe the result of the experiment (Step 1 again), you are startled to see the ball spring to life and scamper off towards some bushes. Your theory was wrong so you devise a new theory: 'It's a small animal' (Step 2 again) and follow it to see if you can find out more. Whether you know it or not you are using the scientific method. You are a scientist. We all use this technique virtually every day. We can't find our keys but think we left them in the pocket of the jacket we were wearing last night, so we go to look – observation, theory, test. We are all scientists and always were. Today, people tend to reserve the word science for particular academic activities to which they give technical names –

astronomy, geology, chemistry, genetics and dozens more – and they call people who are paid to use the scientific method in their jobs 'scientists', but in reality we are all scientists.

Since the 17th century, the word science has acquired an aura of mystique but this is a misunderstanding. Science is not at all mysterious; it's just that triangle. Two things make it seem impenetrable. First, the subjects the professionals study are often very complicated ('How are stars formed?' 'What is inside an atom?' 'How can continents drift across the surface of a solid Earth?') and, second, every branch of science has its own technical jargon which means nothing to us and may leave us feeling excluded and threatened.

When scientists study complicated subjects like star formation, they break the topic down into hundreds of different observations and hundreds of simple theories then test each theory. Sometimes, they need complicated equipment to test their simple theory but ultimately it's only the technology and the overall subject that are complicated. As for the jargon, virtually every branch of human activity has its own vocabulary. Who can understand what a motor mechanic is talking about or name all the tools used by a carpenter? At least in those cases, other motor mechanics and carpenters stand some chance. Science is such a broad subject (or in truth many broad subjects) that few scientists understand what other scientists are saying even within their own discipline. A biologist studying bird classification and a biologist studying bird physiology may as well come from different planets. Neither will understand the other's technical terms, and they are both biologists working with the same group of animals. Science should not be seen as a coherent activity with 'insiders' and 'outsiders'. Most professional scientists find out what is happening in most of science from the newspapers and television, just like the rest of us.

What is not science?

Some subjects don't fall into the category of science because they are impossible to fit into the science triangle. For example, our observations of the world may lead us to theorise that there is a higher spiritual power, a god. We therefore have two limbs of the triangle: an observation and a theory to explain it. The problem arises when we try to devise a test. What experiment would test the theory 'There is a god'? To date no one has ever thought of one. Religion, therefore, is not science.

It has been argued by some people that science is antagonistic to religion and promotes atheism. This is not true. The results of scientific activity do not, and could not, imply there is no god. That would also require evidence from an experiment. As the old saying goes, 'Absence of evidence is not evidence of absence.' Science simply cannot investigate, and hence has nothing to say about, the existence or non-existence of a god. These are matters of faith. A self-professed atheist is just as much a believer as is a bishop. A bishop believes as an act of faith there is a god; an atheist believes as an act of faith there is no god. Science cannot help either of them. Science is necessarily agnostic (from the Greek, 'without the power of knowing'). The only scientific approach to a god's existence is to say, 'I cannot explore that question using the scientific method with any hope of success, therefore I should not try.' Many scientists believe in a god as an act of faith. There is no contradiction here. Science is only capable of investigating the physical universe, yet scientists are human, and two of the cornerstones of the human condition are our logic and our intuition – parallel approaches in the way we build our view of the universe. Science and religion reflect these two approaches. They can survive and flourish together.

Fossils

Applying the scientific method to an investigation of the natural world is not difficult when we are examining the present, but becomes much harder when we start to explore the past (scientists can't record the mating calls of dinosaurs). However, that doesn't mean it's impossible to explore the past using science. As long as a theory is testable, it can be scientific; and a test need not be a laboratory experiment, it can also be a prediction. For example, if birds evolved from reptiles then somewhere in the rocks should be fossils showing a mixture of reptile and bird characteristics. If palaeontologists look for these, they may eventually find them. 'Birds evolved from reptiles' is therefore a testable theory, even though no one knows where to look or how long it may take. In fact, in this example a fossil has already been discovered. *Archaeopteryx*, found in a German quarry in 1861, shows just such a mixture of characters. However, the finding of fossils may rely more on luck than judgement. Most animals and plants are not fossilised when they die, they are eaten or their bodies are destroyed by scavengers or decomposition. Only unusually are some remains protected from destruction and turned to stone. Even then, most fossils will later be destroyed by erosion of the land over geological time or may lie so deeply underground that no one knows anything about them. Fossils are found only when someone interested in fossils stumbles across rocks in the process of being eroded and notices the telltale signs, or when engineers excavate mines or quarries. The chances of any fossil being discovered in the brief period when it's visible are therefore very remote and we shall never have a complete and accurate record of all the animals and plants that lived at a particular time or in a particular place. Studying palaeontology is like trying to analyse a game of football when all you can see are the shadows, and clouds keep drifting across the sun.

Chapter Three

Evolution, Darwin and Natural Selection

Evolution is an old idea; in Europe, it can be traced back to the ancient Greeks more than 2,500 years ago. For centuries, Christians rejected this idea because it contradicted the opening words of the Bible which say God created the Earth and all the species on it, including ourselves, in six days. Here we see the true history of the alleged conflict between science and religion. Science may not be contrary to the idea that there is a god but it is capable of showing the universe was not created in six days.

By the end of the 18th century, with the expansion of European science and with growing numbers of naturalists studying the world around them, the possibility that species might change their appearance over time was increasingly under discussion. In addition to religious resistance, there were two other problems preventing the acceptance of this idea: the length of time such evolution would take and the fact that no one could see a mechanism by which it might occur. At that time the Earth was thought to be only several thousand years old, not long enough for evolution to have had an effect. The realisation around the year 1800 that the complex geological structure of the world had in fact been produced by the incredibly slow action of volcanoes, sedimentation and weathering, just as we see happening today, awoke scientists to the fact that the Earth must be significantly older than they thought. With that, evolution could be taken more seriously.

The timescale involved in evolution is truly beyond the grasp of the human mind. Even today, when we talk glibly about hundreds of millions of years – as I did in the opening lines of this book – our brains are simply not capable of understanding what that means. If non-Christian readers will allow me a biased example, most of us would recognise that the period between ourselves today and Christ 2,000 years ago feels like a very long time. Christ lived in ancient history but if we close our eyes we can probably picture that timescale. If I ask you to imagine 10,000 years ago, you have to imagine five times the period between ourselves and Christ. Now we are truly deep into the past. No event in any history book you have ever read had yet occurred. Our ancestors were still chipping flakes off the sides of pieces of flint and would continue to do so for millennia. This timescale is getting harder to imagine but we still think we can manage it. Now let me ask you to imagine an elapse of time *2,000* times the difference between ourselves and Christ. Such a vast age is almost incomprehensible. It fades far into the distance, well beyond our mental horizon, yet this takes us back only 4 million years. It takes us back to a time when our ancestors were about to leave the trees and walk across the plains of Africa leaving footprints virtually identical to our own. In geological terms this was only a few hours ago. The giant dinosaurs disappeared 65 million years ago, having dominated the Earth for 140 million years before that, and they were newcomers. Life has lived on this planet for more than 3,500 million years. Evolution is very, very slow, but there has never been any rush.

Darwin

In 1859, Charles Darwin published his book *On the Origin of Species by Means of Natural Selection*. Natural selection was the mechanism Darwin proposed to explain how evolution could work. In his book, Darwin made several critical observations: resources in nature (such as food or living space) are limited; there is competition for them; and within each species individuals are slightly different from each other.

He argued that in a competition in which the competitors show different characteristics (one leopard being able to run slightly faster than another; one mouse having slightly paler fur than another) some characteristics would give an advantage and some would not. In a struggle for life – as he put it – the characteristics giving an advantage could lead to their owner, and hence the characteristic, surviving. In this way, some characters would automatically be selected by nature, survive and be inherited by the next generation. During this process of natural selection, with some characters passing from one generation to the next and others being eliminated, the species showing the characters would change over time, there would be evolution.

Darwin is often said to be the father of evolution. In fact, he was the father of natural selection, which is how evolution works. His ideas have been tested extensively since his book appeared and his theory has long since ceased to be a theory. Evolution and natural selection are now established as facts.

EVOLUTION, DARWIN AND NATURAL SELECTION

Natural selection

Natural selection changes the average appearance of a species, not individual animals. To oversimplify: imagine a herd of gazelles in which each gazelle has legs slightly different in length to the others in the herd. There are therefore tall gazelles and there are short gazelles (this could equally be a room full of people of different heights, but what happens next would not bear thinking about). If all the gazelles with the shortest legs are caught and eaten by lions because they can't run fast enough, only the gazelles with longer legs will survive. The short-legged gazelles will not have lived long enough to reach breeding age so only long-legged gazelles will produce young. *All* the next generation of gazelles in this herd will therefore tend to have long legs. No individual gazelle has increased the length of its legs but the average length of legs in the herd has increased. There has been evolution. Evolution therefore acts at the level of reproduction; the natural selection of one generation affects the appearance of the next generation.

Natural selection of behaviour

Some characters selected by nature may be behaviours, not features governed by our genes. A group of animals – including early humans – that decides to drink at the same water-hole as predators, at the same time as the predators drink, and who then jostle for position amongst themselves rather than remaining vigilant, are unlikely to last long enough to teach this behaviour (deliberately or by example) to their children. Indeed, they are unlikely to last long enough to have children. On the other hand, a group that allows the predators to drink and disperse before they themselves journey to the water-hole, and who then post lookouts, may survive to pass this behaviour to the next generation.

To add to the complications, some features may be behaviours *and* governed by our genes. They are inherited behaviours not learned behaviours. Examples of this are the newborn baby's innate abilities to suckle and to cry, and its grasping response. A very young baby will grip a finger firmly in their fist, and can even support their own bodyweight, long before they have had an opportunity to learn to do this. When we watch other primates carrying their newborn infants on their backs, the infants' hands grasping their mothers' fur, it is not difficult to see the origins of this behaviour or how it came to be a behaviour of all newborn humans.

The virtually universal human fear of the dark may also fall into the category of inherited behaviour. Hundreds of thousands of years ago when people

EVOLUTION, DARWIN AND NATURAL SELECTION

lived in the open air surrounded by predators, this would be a very advantageous fear. It would not have been a good survival strategy to walk around in the middle of the night when you couldn't see what was around you, or to stumble into dark caves without a light. Individuals who feared the dark and stayed in a protected place after dusk would be more likely to survive the night and pass this fear (if it is in the genes) to their children, and ultimately to us. Nowadays, in most cultures our homes are not dangerous places when the sun goes down but our natural fear of the dark remains, and has been exploited by virtually every horror film ever made.

Survival of the fittest

'Survival of the fittest' is a phrase often heard when people talk about evolution. This does not mean survival of the most healthy, it means survival of those animals or plants which are best fitted to their environment. In the lion vs. gazelle example above, it was the long-legged gazelles that survived because they were the gazelles in which there was the best agreement – the best fit – between their bodies and the needs of survival.

Sometimes in history, parts of animal populations have survived not because they were the best fitted for survival but because something happened to other members of the species and only their group was left alive to pass on its genes. This is more a case of 'survival of the luckiest'. This happened in the northern Pacific Ocean to the population of Northern Elephant Seals. In the 19th century, man hunted this species almost to extinction and by 1890 there were fewer than 20 individuals left. This handful of animals did not have some adaptation that made them harder to hunt; they were simply the last animals to be slaughtered.

In fact, they were not slaughtered and, with restrictions on hunting, their offspring now number more than 30,000 individuals. However, all the genes the species contains now come from fewer than 20 animals and today there is much less genetic variation than there originally was. It is as though the genes in this species have been forced through the constriction in an hourglass and most genes did not survive the journey. The raw materials for natural selection have been severely reduced and the evolution of the species will certainly be affected.

In Africa, cheetahs seem to have passed through a similar bottleneck several thousand years ago. There is so little genetic variation in modern cheetahs that for some reason the population must have been reduced to only a few individuals.

To summarise: natural selection (and sometimes

EVOLUTION, DARWIN AND NATURAL SELECTION

catastrophes) effectively edit the individuals of a generation. Natural selection removes some before they can reproduce; it inhibits the reproduction of others and enhances the reproduction of some. In this way, the next generation only inherits edited characters and these edited characters change the appearance and functioning of the species. As many of these characters are governed by our genes, it may be useful to think briefly about what genes are.

Chapter Four

Genes

'Gene' is from the ancient Greek word for descent, or birth (from the same root as genealogy or genesis). Genes are inherited instructions elling the body how to build and maintain itself. These instructions may be for something internal like the production of enzymes in the intestine or some-thing obvious like height or the shape of our nose.

Each gene is a short string of molecules. These are attached end to end to form threads of DNA or de-oxyribonucleic acid – so called because it's an acidic molecule found in the cell nucleus (a 'nucleic acid') and includes the sugar Ribose from which an oxygen atom has been removed ('de-oxy ribo').

YOUR BODY

These threads of DNA are found in the nucleus of most cells and the human body contains approximately one hundred million million cells (100,000,000,000,000 or 10^{14} cells), each of which contains the full set of genes for making and running the whole body. A cell in the eye therefore contains the genes for making a stomach or a kneecap, even though they will never be used. This is like every library in the world containing a street atlas of every town in the world, even though most people only consult maps of their local area.

In humans, each nucleus has 46 threads of DNA, together more than 2 metres in length and containing about 24,000 genes. These 46 DNA threads are arranged in pairs. This is because we all receive one of each pair from each parent; 23 from the mother's egg and 23 from the father's sperm. This is a bit like getting a pair of socks for your birthday but your mother gives you one sock and your father gives you the other, except in this case you get 23 odd socks from each of them. When the gifts are put together, they make 23 matched pairs.

Each thread of the pair contains genes for features of our body – hair colour, eye colour, arm length – but the other thread of that pair also contains genes for those same features. The threads are twins. This means we all inherit two genes for most features, not

GENES

one, but how this works in practice is not something we need to consider in this book.

At certain times during the life of a cell, each DNA thread twists in upon itself like a tangled rope to form a shorter fatter structure which early naturalists called a chromosome (Greek for 'coloured body' because, in the early days of the microscope when people were staining cells with different dyes to make them visible, the chromosomes sometimes appeared as short dark ribbons. Today, we call the threads chromosomes whether they are densely twisted or not). These twisted 'coloured bodies' appear when the cell is about to divide during tissue growth, wound healing or cell replacement.

Most cell types are replaced continuously. Skin wears away all the time, with new skin being formed underneath it, and red blood cells (which carry oxygen around the body) only have a life of about 120 days. Each of us has billions upon billions of red blood cells (one large drop of blood contains about 500 million) and 170 thousand million (170,000,000,000) new red blood cells are created *every day* by your bone marrow to replace those being destroyed by – or perhaps we should say recycled by – your spleen, liver and (again) bone marrow. When a blood donor donates half a litre of blood, the body loses about two and a half million million (2,500,000,000,000) red cells and it takes about 50 days to replace them. (It doesn't take 15 days as would be expected from the multiplication:

170,000,000,000 per day x 15 = 2,550,000,000,000

because these cells are lost to the body and cannot be recycled. Donors must get new raw materials from their food before they can make new cells.)

Most animals have a nucleus containing DNA in their red blood cells, as in the other cells of their body, but mammals, including ourselves, lose the nucleus of the red blood cell as the cell is formed. This is how forensic scientists can tell whether a bloodstain came from a person or from the preparation of someone's chicken supper.

As cells in the body multiply to produce new skin or blood, or any other tissues, the chromosomes, and hence the genes, must be copied into each new cell. Genes are copied into an egg when it forms in a woman's ovary, or into a spermatozoon when it's produced in a man's testes (testicles). They are copied when a cell divides to become two cells, as when a fertilised egg begins to grow in the womb. By the time a baby is fully formed, all its genes have been copied numerous times.

Mutations

In science fiction stories, mutants are invariably bad things. In the real world, 'mutation' just means

GENES

'change'. Genes are copied during growth and as they pass from generation to generation, and whenever anything is copied accidental changes can creep in. An analogy can be made with an old story about the First World War. In this tall tale, the officers in the front line send a message back to headquarters saying: 'Send reinforcements, we are going to advance.' The message is not written down but is passed by word of mouth from one trench to the next until it reaches its destination. Unfortunately, by the time it arrives it has mutated into: 'Send three and four pence, we are going to a dance.' What is noticeable about this is that the message did not deteriorate into complete gibberish. It remains a perfectly logical message: 'Send three shillings and four pence', but now it has no relevance to the situation at hand and no relation to the original message sent. The problem arose because the message, like genes, was repeatedly copied.

Mutation of a gene may stop it working (in an egg, sperm or embryo, if the gene is important enough, this may kill the embryo) or it may not have a serious effect, or it may even make the gene more effective. As the change is a random accident, the results can be very variable. Like the message in the trenches, the gene may not be destroyed or garbled completely but it may cease to do the original job. Whether it survives in the population in its changed form will depend on whether this new form is harmful to its owner. A mutated gene that kills its owner while he or she is still in the womb will die with them. It will never be inherited by anyone else and will disappear from the world immediately. On the other hand, some mutations may be beneficial and will spread quickly in the population.

Here is an another analogy. Genes are instructions. They are like a cake recipe where the body is the cake. Imagine a rather uninspiring cake recipe that includes the instruction to add 100 grams of coconut. Not everyone likes coconut. Some people make this cake and some eat it but very few ask for a copy of the recipe. One day when someone does ask and is transcribing it into their notebook, they have difficulty reading the original handwriting. Instead of 'coconut', they think it says 'chocolate'. They make the cake and instead of coconut add 100 grams of chocolate. The cake is delicious. Everyone who tastes it asks for the recipe. Suddenly instead of there being five copies of the recipe in the world, there are 500, then 5,000 and soon there are 5 million. The copying error, the mutation, has been very successful for this recipe and the cakes are now everywhere. This can happen to genes too. This can even happen to genes without the change being obviously beneficial. A mutation has occurred in blood which does not appear to have any benefit at all, yet like the cake recipe it has been very successful; for most of us it determines which blood group we are.

Blood groups

On the surface of each red blood cell are molecules called antigens. These are of two types: Type A and Type B. The type we have in our blood is inherited from our parents and this gives us our blood group.

Everyone talks about blood groups using three letters A, B and O. This is the convention, but in reality it is wrong. The 'O' here is not the letter O, it is the number zero, used to indicate that the A and B antigens are absent.

In the early history of our species, each blood group gene produced either A or B. If a child inherited an 'A' gene from their mother and an 'A' gene from their father, they would be blood group A. If they inherited two 'B' genes, they would be group B. If they inherited one of each, they would be group AB. However, somewhere in the past, one of these genes mutated in such a way that it failed to produce either antigen. Although this would not appear to have conferred any advantage to its owner, this mutation has now spread throughout the human species until today it is the most common form of the gene. The range of blood groups is now therefore far greater than it used to be.

Today, if we get a gene from each of our parents that creates antigen A, we will still have two A genes ('AA') and will be blood group A. On the other hand, if we get a gene from one parent that produces 'A' and a gene from the other parent that produces nothing, we will have the genes 'A0' (A + zero) but when our blood is tested we will still show a positive test for antigen A and will still be blood group A. There are therefore now two ways of being blood group A (three if you count as different options getting the single A from the mother or getting the single A from the father – 'AA', 'A0' or '0A'). The same applies for blood group B. It is still possible to get an A from one parent and a B from the other, when we will be group AB (now quite rare), but if we get the gene for nothing from one parent and the gene for nothing from the other parent ('00') we will produce neither antigen and will be blood group O (which despite its meaning is still pronounced as the letter O, 'oh'). There is therefore only one way of being group O (by being '00') but this is the most common blood group. This is because most parents who are group A are 'A0 or 0A' not 'AA' and most who are group B are 'B0 or 0B' not 'BB', so adding these to the parents who are '00' means genes producing neither antigen are the most common.

The complication caused by this mutated gene is

that it is now possible for children to have blood groups quite different from either of their parents. For example, both parents can be group A and the child can be group O,

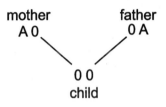

or one parent can be A and the other B, yet the child can be AB or O.

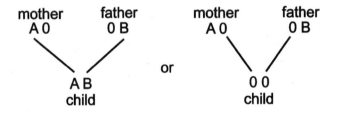

The child in the last example could also be group A ('A0') or group B ('0B') but they could never be 'AA' or 'BB'. Which of the possibilities becomes reality is pure chance. It depends which of the mother's genes occurs in the egg that is fertilised, her A gene or her 0 gene, and which gene occurs in the father's successful sperm, his B gene or his 0 gene (only one chromosome of each pair goes into an egg or sperm).

If the parents above had more than one child, the children could also be different from each other.

The percentages of the different blood groups in today's British population is [with the percentages for the United States in brackets] – Group O: 45% [45%]; Group A: 43 [40]; Group B: 9 [11]; Group AB: 3 [4]. However, both the UK and USA have populations composed mainly of immigrants. (Those people in the UK who complain about immigration undermining traditional Anglo-Saxon culture tend to forget that the Angles and the Saxons *were* immigrants. Without immigration, there would be no Anglo-Saxon culture. England takes its very name, not from its native peoples but from its immigrants: Angle-land.) In the immigrant populations of the UK and USA, the genes for blood groups will therefore be quite mixed and will not reflect the distribution of blood groups in the original Celtic inhabitants of the British Isles or in the native North Americans.

Native North Americans had virtually no blood group B in their original populations and this has led to suggestions that they are all descended from one small group of individuals who migrated to North America from Asia across the Bering Strait at the time of the last Ice Age – a small group who just happened to have no blood group B among their members. This creates a situation like a survivor population (as noted for Northern Elephant Seals earlier) but as in this case the rest of humanity had not become extinct; scientists call this type of bottleneck, where one group descends from a small number of breakaway founders, a 'founder population'.

Teams of genes

The situation with blood groups is complicated enough, but for most of our genes it gets worse. It may be that few features are controlled by just one gene (or correctly one gene pair) which we can conveniently label as an 'eye-colour gene' or a 'height gene'. For most characteristics, it seems many genes contribute to the final result, and they do this by interacting with each other. Eye colour is now known to be influenced by at least three different gene pairs, although there is a suspicion many more are involved. Skin colour is also controlled by more than one gene and this will probably be true for most aspects of our body.

Nor does the complexity stop there. It would be wrong to see the body simply as a machine, programmed by genes and built piece by piece like bolting prefabricated elements together. In biological systems such as ourselves, the parts are not stiff and dead like the components of an engine. Not only do the genes collaborate to create the elements for our body, but during our development as embryos these elements also interact with each other and affect how each develops. Skin may develop as skin not just because it has genes telling it to become skin but because neighbouring cells confirm this instruction and tell it what sort of skin to become. If it is on the scalp, it may become skin with hair. If it is in our mouths, it may become thin, hairless skin with glands to keep it moist. Both types of skin have the same genes; virtually every cell in our body has the same genes. We should not assume the genes in a cell inside the mouth and the genes in a cell on the back of the head know where they are in the body. Many tissues may develop as they do, not because of the genes they contain but because of the other tissues that surround them. Our bodies are more than just the products of an assembly line, they are the result of an integrated community of cells and tissues growing and communicating throughout our lives, and never more so than when we are still in the womb.

With such layers of complexity contributing to the final result, the current attempts in some laboratories to locate one 'baldness gene' or one 'homosexuality gene' (both reported in the media in recent years) are almost certainly wild goose chases.

Chapter Five

Evolution in Practice

The body shape of any one species at any one time is a collection of features its history has handed down to it. If a species is to survive a change in its environment or to exploit a new way of life, natural selection can only act on the body that exists. It cannot always address a species' problems in the most efficient way because there may not be the genes for this. If the species is to survive, natural selection must work with the tools history has given it. It's time for another analogy.

If your ship sank without warning and you were washed alone onto a desert island, you would have to survive with whatever you had in your pockets at the time. The best survival aids for the situation might be a 36-piece set of woodworking tools and a copy of *Island Survival* (4th edn) by R. Crusoe, but you don't have these. You would have to use what you did have as best you could and adapt or die.

In your pocket you might have a coin. You sharpen the edge on a stone and use it to whittle points onto arrows for hunting and fishing. A knife would have been better but you don't have one. The coin will do the job and, in fact, the raised lettering that indicates the denomination helps your grip when cutting. Later you find a better way of sharpening your arrows. You no longer need the coin as a cutting tool but you do need a weight for your fishing line. The ideal solution would be a split lead shot but you don't have any lead. Instead, you bend the coin double and clamp it over the line. Now the raised lettering aids the grip of the coin on the line. You have found two important uses for something that was initially for a completely different purpose; a purpose of no relevance whatever to your current situation. The coin was not ideal for either job but it worked. You catch fish and you survive. Meanwhile, the coin bears the scars of its past history. It still has the lettering and a mainly curved edge from its life as a coin, and the sharp edge from its life as a cutting tool. The lettering is now useful for a reason never intended when it was first developed; the

GENES

curved and sharpened edges serve no purpose in a weight, but do no harm.

What we do by craft, evolution does as an automatic consequence of natural selection, but the results can be similar. Like the coin and its lettering, the bodies of many animals include structures now employed for purposes that were not their original use. The teeth of sharks are modified skin scales; the wings of birds are modified arms; the membranes of bats' wings evolved as skin to cover the body but later extended to provide a large aerodynamic surface. Also, like the curved and sharp edges of the fishing weight, animals' bodies contain structures which serve no purpose today but which were evolved by ancestors who did need them. These have passed down through the generations ever since. You are probably sitting on the remnants of a tail, a tail evolved and used by your ancestors but now just a line of bones inside your body at the bottom of your spine. Another example would be the dewclaw of dogs. This is the small redundant toe which is seen above the foot and doesn't touch the ground. The ancestors of dogs had five full toes but, as the dog evolved to move faster, one toe shrank and moved up the leg away from the others (reduction in toe number is common in animals which have evolved to move quickly over the ground – see p127). Dogs now have only four functional toes. Dewclaws serve no purpose today and dog breeders often have them surgically removed. If the dog's ancestors had remained subject to the pressures of natural selection, rather than forming a relationship with humans and allowing us to control the evolution of their bodies, it is possible the dew claw would eventually have disappeared completely, or at least become wholly internal, like our tails.

Evolution is not a perfectionist

Once a feature has been lost in evolution, it is almost impossible to resurrect it. Natural selection only acts on what is visible to it so it is more usual for evolution to provide solutions to new problems by modifying parts of the body currently in use. For example, birds evolved from early reptiles which evolved from early amphibians which, in turn, evolved from early fish. Early fish, amphibians and reptiles all had long tails and these were originally used for swimming. In their evolutionary journey into the air, birds lost their heavy, bony tails and replaced them with long, light tail-feathers. Only the stump of the 'parson's nose' remains. When some birds returned to a watery existence and began to swim again (penguins for example), they did not re-evolve a long swimming tail, they used the limbs that propelled them when they flew, their wings, and now fly underwater. In response to this, the bones of the wing have become stronger and heavier than those of other birds to cope with the density of water, which is much harder to push through than air, and as a result penguins have lost the ability to fly in air. Penguins also float and swim on the surface of the sea but again a swimming tail did not re-evolve. Here the hind feet became webbed and push the bird forward as they do when it walks on land. The actions of paddling are very like those of walking.

The ideal engineering solution to the problem of how to propel a bird underwater might have been to give it a fish-like tail, but evolution does not plan and design bodies. Evolution is what happens because there is selection; it is not why there is selection. The penguin adapted to its new habits and survived because each generation was edited by natural selection. Small incremental changes modified the average appearance of penguins in such a way that they survived as underwater fish hunters. Nature does not demand the perfect solution, only a solution that works.

This gives us the answer to an age-old question that has plagued many mothers: 'Why is giving birth so painful?' The brutal explanation is, 'Because it doesn't have to be painless.' Evolution doesn't care whether birth is agony or ecstasy as long as it is successful. As long as healthy babies continue to be born, the accompanying pain has no relevance to natural selection, regardless of its relevance to the mother. The current degree of childbirth pain is probably a relatively recent phenomenon, increasing only over the last few million years as human brains have increased in size. With the brains of babies in the womb enlarging relative to their ancestors, a larger head has had to pass through the same size of birth canal. As this has continued to happen successfully there has been no drive for natural selection to increase the size of the canal. The result is pain.

'Nature abhors a vacuum'

This is a phrase often used by biologists, but its meaning is not immediately obvious. As Darwin noted, species compete for resources. However, it's always easier to make a living if you have no competitors, so, if an opportunity exists in nature which is not already being exploited, something will invariably evolve to exploit it.

This is very like our commercial world today. If workers in offices are finding it hard to leave the building to buy lunch, someone will eventually set up a business making sandwiches and taking them to the offices to sell to the deskbound inmates. These sandwich entrepreneurs make their money by

GENES

exploiting a gap in the market – you can probably think of dozens of similar examples. In fact, commerce is a good analogy for nature. This 'gap in the market' is the commercial 'vacuum' everyone is looking for, something they can do which no one else has thought of, where they will have a monopoly with no competitors. They can make a living while expending the minimum of effort. When others start to compete with them, they will usually try to modify their business to become more efficient (sell sandwiches that cost less to make), or provide an improved service (sell drinks as well), or branch out into new areas where they will again have no competition (sell sandwiches at construction sites), or simply try to out-compete the competition. This is Darwin's struggle for life (in this case commercial life). Businesses therefore evolve in the same way as species evolve.

In nature, many species of bird feed on flying insects – they are a protein-rich diet – but there are also insects flying at night when birds are sleeping. Nature abhors a vacuum so this gap in the market was exploited by some nocturnal mammals which evolved wings and became bats. They even evolved wings from their front legs, just as the birds had done before them. Bats superficially resemble birds because they live in the same way.

YOUR BODY

Businesses that do the same thing will also tend to look like one another, even if they have no direct connection. This is forced upon them by the requirements of their activity. Sandwich providers on opposite sides of the world must share common features: they will use bread and sandwich fillings; they must have a facility where the sandwiches are made; they must have people making sandwiches; they must have transport to enable them to reach the offices where their customers are; they must have baskets or trolleys to carry the sandwiches; and they must handle money. These are constraints imposed upon them by their lifestyle. Nature is no different.

Animals that exploit the same gaps in the market in different parts of the world (biologists call these gaps 'niches') tend to resemble each other. Sea-birds in the southern hemisphere that paddle on the surface of the sea and dive for fish, swimming underwater by flapping their wings (penguins) look like sea-birds in the northern hemisphere that paddle on the surface of the sea and dive for fish, swimming underwater by flapping their wings (auks like the guillemot and razorbill). Mammals that live permanently in the water hunting fish at high speeds (dolphins) look like predatory fish that live the same way, not like other mammals. The extinct reptile ichthyosaurs also looked like dolphins and predatory fish because they shared this lifestyle – even their name means 'fish lizards'.

Different animal groups can therefore end up mimicking each other's bodies because they live in similar ways, even though they are not closely related. Biologists call this 'convergent evolution' because different animals have converged on the same body shape independently, rather than inheriting the same shape from a shared ancestor.

Evolution never sleeps

Evolution never stops, but nor is it ever heading towards a goal; it is simply a passive response to natural selection, which in turn never stops. Humans are not, as was once thought, the pinnacle of evolution; evolution has no pinnacle, it goes on for ever. We are not even today's most advanced species. Every species alive has been evolving for the same length of time – since life first appeared on the planet – and all deserve equal consideration as successful survivors.

There is a tendency to look at body shapes that have existed in the fossil record virtually unchanged for tens of millions of years – for example, ferns, sharks and crocodiles – and see them as primitive forms frozen in time. However, evolution does not just affect the outside of an animal or plant, it also affects the inside, both the organs and how they function. Even the chemistry of all living things is subject to change. Just because a species looks similar on the outside to a related species that lived millions of years ago does not mean that that group has not evolved for millions of years. Any gardener will tell you ferns suffer from very few diseases and are attacked by very few pests, and very few animals eat them. Their chemical defences are extremely efficient. That does not mean the ancient ferns had similar advantages. Ferns may have spent millions of years evolving these defences while their outer form changed very little. Sharks also suffer from few diseases and crocodiles have such an effective healing capability – recovering from severe wounds in very polluted waters with little infection – that medical researchers are now trying to identify the factors in crocodiles' blood that defend them against bacteria in the hope of discovering a new drug similar to penicillin.

Nor are species alive today somehow better than species in the past. Each is or was tailor-made for its own environment. Past species were not intermediates on the way to something else. To put this in a context, we can look back to people who lived 10,000 years ago and see they had less knowledge, less technology, less medicine and less comfort than we have today, but, before we describe them as primitive or backward, we should consider what the world will be like for people 10,000 years in the future. They will have lives and facilities incomprehensible to us. Do we want them to look back at us and see us as primitive or backward intermediates whose only purpose was as stepping-stones on a path leading to them? I don't think so. We do not exist to prepare the way for future generations. We react to, and survive as best we can in, the world of the present. Evolution does the same. Every species is right for its time. Its time may be short or it may be long, but the very fact that it existed at all indicates that at that moment it was a survivor. We should remember it took life 3,500 million years to produce the dodo. This bird was honed by natural selection to fit the environment of Mauritius. It was only when Europeans arrived and changed that environment by introducing European species that the dodo found itself ill equipped, just as we would be if someone released a pride of lions in our local shopping centre (dodos nested on the ground and were probably too vulnerable to the egg-raiding pigs introduced by the sailors).

Without human interference, the dodo was a survivor, but survivors are not always easy to spot. No species that lived on the planet 400 million years ago still lives on it today, yet if they were all extinct there would be no life today. There are two ways for species to disappear: their representatives may dwindle in number until the last one finally expires — the classic extinction — or the species may evolve into one or more other species, unrecognisable as the same animal or plant as its ancestors. About 4 million years ago, just such a species swung down out of the trees and strode out onto the grasslands of Africa. That species no longer lives — if we passed it in the street it would be instantly recognisable as different to anything we know — but nor is that species extinct. Like so many before, it has survived by shape-shifting down the ages and into today's world. Now it is time to consider its descendants.

Chapter six

When We Were Fish

Modern shape-shifters such as ourselves have evolved from the first vertebrates. These were jaw-less filter-feeding fish which appeared in the sea more than 500 million years ago. There are still some jawless fish today: marine hag-fishes and lampreys (like eels but with circular sucker-like mouths) but these bear little resemblance to their early jawless ancestors.

The stiffening rod down the back of these first vertebrates was not bone but a tube of tough membrane filled with a fluid under pressure. The first fish with bones did not appear until 100 million years later. Today, fish with bone skeletons are called 'bony fish', even by scientists, to distinguish them from

YOUR BODY

Pteraspis, an early jawless fish which lived more than 400 million years ago (about 25cm long)

sharks and rays which have a skeleton made primarily of cartilage – a solid but softer substance which a carnivore would call gristle. Most fish in the world today are bony fish.

Before fish arrived with their internal stiffening rod, invertebrates had evolved a number of different ways to give support to their bodies. Segmented worms used hydraulic pressure controlled by layers of muscles in the skin. Crustaceans had a hard skeleton on the outside, like a suit of armour, moved by muscles attached to its inside. The fish had their skeleton on the inside with muscles packed around the stiffening rods.

Muscles are simply blocks of cells that have evolved to contract. All they do is shorten, which pulls closer together whatever is attached to their ends. They cannot push. In this sense, they are like ropes, which can pull heavy loads but can't push anything. If muscles in a vertebrate are to move part of the skeleton more than once they must operate in pairs: one to pull the rod one way, then another on the

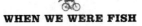
WHEN WE WERE FISH

other side to pull it back again. When the second muscle contracts, the first muscle relaxes and is stretched ready for its next contraction. This arrangement would develop further as the vertebrates evolved into new forms — especially when they left the water — but in the early fish the main role of the muscles was to produce the characteristic S-shaped flexing of the body that resulted in swimming.

A fish moves by bending its body from side to side in such a way that a wave passes down its flank, pushing on the water as it goes. This is like shaking a loose rope from side to side to make waves pass along it. To produce this movement, the fish's spine bends as the muscles of each side pull alternately. The weight of the fish's body is supported by the water so the only function of the spine is to prevent the whole fish shrinking into a short fat horseshoe when the muscles of one side contract. At this stage in our evolution, our spine was therefore relatively weak and acted only like a flexible tent-pole, keeping the body extended and converting the work of the muscles into gentle undulations. This wave alone moved the fish forward but it was neither particularly efficient nor particularly stable. Fish soon developed new structures that increased their ability to push against the water and stopped them toppling out of alignment as they moved. These were the fins, some of which would later become our arms and legs.

Fins

Early jawless fish had fins along their mid-line and scientists believe some also had a long fold of skin along each side of their body. With time, these folds shrank, leaving only one pair of flaps towards the front and another pair towards the back.

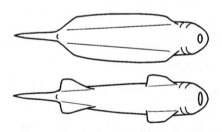

A representation of early jawless fish from below, showing how folds of skin could have become modified as paired fins

WHEN WE WERE FISH

Even today, modern fish have various arrangements of fins along their body's mid-line – above, below and at the tail – but only ever have two pairs of paired fins, the pectoral fins at the front and the pelvic fins at the back (although in some species the pelvic fins have migrated forwards and lie below and in front of the pectoral fins). The paired fins are used by today's bony fish as brakes and for slow manoeuvring. It is because these longitudinal folds of skin were reduced to two pairs of flaps that all other vertebrates, including ourselves, now have two pairs of legs, a pair at the front and a pair at the back. Our limbs evolved from these paired fins and, when we are small embryos in the womb, little more than 5mm long, our arms and legs begin their growth as two small semicircular flaps on each side of our body. If the folds had originally formed three pairs of fins we would have six limbs today, but two pairs were all that was required to stabilise the early fish – one pair in front of the centre of gravity and one pair behind it.

Jaws

By the time the pectoral and pelvic fins had evolved, natural selection had also changed the front of many fish. Inside the head, which was encased in bony outer plates as the head of the earlier fish had been, jaws had begun to develop. The earliest fish had a series of gill slits just behind each side of their heads, as sharks have today. Between the slits the skin was strengthened by rods of the skeleton called gill arches. Over time, the first gill arch of each side began to fold forward around the simple opening of the mouth. In doing this, it was able to alter the shape of that opening and the first jaw had appeared.

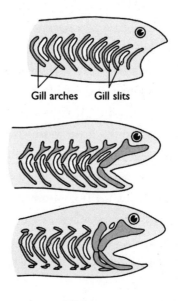

One view of how gill arches may have formed jaws

With this, the fish could feed on larger, more solid foods which they could now grasp in their mouth. It was presumably the advantage of being able to feed in this way, rather than merely filtering minute particles

WHEN WE WERE FISH

from the water, that led to this change. With the arrival of a grasping mouth came another innovation to make this even more successful. Part of the bony head plates around the mouth opening began to develop a rough edge for gripping the food. Natural selection worked on this and eventually the new mouth acquired teeth.

Fins to limbs

In modern bony fish, the muscles controlling the paired fins are inside the body. The projecting fins are only a thin membrane supported by fine struts. We, on the other hand, are descended from an ancient type of bony fish which had the muscles controlling its fins in a fleshy lobe jutting out from the body, carrying the fin membrane on its margin. These lobe-finned fish are represented today only by rare survivors such as the coelacanths of the Indian Ocean and some tropical freshwater lungfish, but they were common in freshwater habitats until about 255 million years ago.

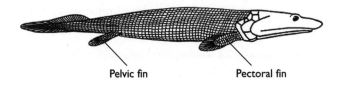

The lobe-finned fish Panderichthys (length about 1 metre)

YOUR BODY

Biologists don't know why lobed fins like this evolved. As some species of fish left the sea and began to colonise rivers, lakes and swamps, they would have encountered more areas of shallow water than their ancestors in the oceans. Fleshy lobed fins might have helped them move across the bottom of shallow, plant-infested waterways. These muscular pegs projecting from the lower sides of the body would catch more readily on rocks or weeds and would lever the fish forward during swimming. At low speeds, this would aid the push from its tail.

Whatever the reason for the appearance of lobed fins, this modification for freshwater life seems to have been very successful. In some groups, the end of the lobes lost their soft feathery margin and instead developed a fringe of bony rays. These may have been more efficient at snagging gravel and plants, or they may have allowed the fish to expand the end of the lobe to give more push against the water. Whatever the reason, these bony rays would later become fingers and toes.

As these ancient fish moved through the lakes and swamps with their primitive muscular 'limbs', they were becoming increasingly specialised for living in a shallow freshwater environment, but this environment contained hazards their ancestors had never encountered in the sea.

WHEN WE WERE FISH
Lungs

Shallow bodies of freshwater, especially ponds, can lose oxygen catastrophically during hot weather. Even at 15°C, water only holds one-thirtieth as much oxygen as the atmosphere. Today, many freshwater fish respond to falling oxygen levels by swimming very close to the water's surface where oxygen dissolves from above and levels are higher. From this behaviour, especially in fish that feed on insects floating on the surface and consequently gulp in air when they feed, it is a short step to gulping air directly for its oxygen, as long as the fish has a way of extracting this oxygen.

Among today's fish, air-breathing organs have evolved several times in different groups which live in bodies of freshwater with seasonal slumps in oxygen levels. Examples are the electric eel *Electrophorus*, which gulps air into its mouth where the oxygen is absorbed through the thin skin, and the catfish *Hoplosternum*, which gulps air into its intestine where oxygen is absorbed by special patches of fine blood vessels in the intestinal lining. It is not difficult to imagine primitive fish hundreds of millions of years ago living in similar environments gulping air for the same reasons. The more efficient they could be at extracting the oxygen, the less dependent they would be on the unreliable water for breathing, and the more likely they would be to survive. Some ancient fish even developed specialised organs for absorbing oxygen from air. We call these lungs.

In the embryos of air-breathing vertebrates today, including ourselves, the lungs develop as pockets growing out of the wall of the embryo's intestine, which suggests lungs may have developed in ancestral fish with air-gulping habits similar to those of *Hoplosternum*. This also explains why we eat and breathe through the same opening, our mouth, and why our windpipe branches off our gullet somewhere in our throat. (We can also breathe through our nose but that is just a tube leading into the back of our mouth.)

The move to dry land

Fish first moved from water to land more than 360 million years ago, a time of occasional droughts and floods when evolution was creating many new fish types.

No one knows why fish first ventured onto land but many of today's vertebrates successfully exploit another medium. Mammals may spend much of their time in water (otters, beavers) and some may become adapted for an aquatic life (seals and sea lions) or even permanently aquatic (whales, dolphins, dugongs, manatees). Mammals may feed in the air and become adapted for an aerial existence (bats). Birds may feed in the water (ducks, swans, pelicans) and some may be adapted for an aquatic life (penguins), although no bird lives permanently in water. Most birds feed on the ground and some have completely lost the power of flight (ostriches, rheas, kiwis). Many living reptiles have become adapted for an aquatic life (crocodiles, marine iguanas) and some spend almost their whole lives in water (turtles, terrapins, sea snakes). In prehistoric times, some reptiles were fully marine (ichthyosaurs, plesiosaurs) or airborne (pterosaurs). Invertebrates too have made numerous transitions. Snails, crustaceans and worms are all found in the oceans and in freshwater and on land. We should not be surprised that some early fish founded a successful frontier lifestyle on the riverbank. For animals that could already breathe air and were already walking underwater, it was literally a short step.

Early fish may have moved onto the land in pursuit of food. The shape of their teeth suggests they were not herbivores, and if they were feeding on insects and other arthropods in the shallows it may have been a successful strategy to pursue them further and further on to the bank. Some killer whales (Orca) today will drive themselves through the surf onto beaches to catch sea lion pups, even though they may have great difficulty getting themselves back into the water. If this activity proves very successful, it is possible to imagine the descendants of these whales developing stronger and more flexible front fins to help the process until eventually they evolve into a species with legs which hunts on beaches. Such whales could eventually become permanent land animals (as their ancestors once were).

Early fish may also have found sanctuary from predators in the shallows or on sandbanks. The modern mud-skipper, a small marine fish with strong pectoral fins which can move on mudflats exposed to the air, leaves the sea both to feed and to escape predation.

On land, early fish probably still travelled by fish-like undulations of the body with the primitive limbs acting as pegs which would catch the ground at pivot points and lever the animal forward, but, although they didn't know it, they were in the first stages of evolving into true tetrapods (Greek for 'four legs' – the Latin 'quadruped' means the same thing).

Ears

What we hear as sound is the vibration of the medium around us. (Despite the best efforts of science-fiction films, explosions and rocket engines in outer space are silent; there is nothing in the vacuum of space to transmit any sound.) In fish – including our ancestors – the body is of almost the same density as the surrounding water. Sound underwater

WHEN WE WERE FISH

therefore passes straight into a fish and no sophisticated organs of hearing are needed. Today's fish have a small internal structure for detecting vibrations but no external ear. When some early fish began to spend more time in the air, they would not have been able to hear well because the density of the air was much less than their tissues and sound waves pass only poorly from air into bodies. They would have been able to sense some vibrations through the earth, but as they were the first land vertebrates there was not much on the land large enough to make a vibration. A rock-avalanche or a large falling plant would be heavy enough to vibrate the ground, but by the time it did that it would be too late for the poor fish to get out of the way. They may, however, have been able to detect the movement of other fish and this would have had some advantage for finding prey or avoiding predators. It would be a long time before an ear evolved that was fully adapted to airborne sounds, but these walking, air-breathing fish probably still spent most of their time in the water.

What we inherit from our fish ancestors

Our time as fish has left us with one central spine, two arms, two legs, jaws, teeth, lungs and the habit of both eating and breathing through our mouth.

Chapter Seven

When We Were Amphibians

As time passed, the air-breathing fish became increasingly adapted to life out of the water. By 360 million years ago, some had changed sufficiently to be called primitive amphibians (from the Greek meaning 'living on both sides'), although they were like no amphibian alive today. In overall shape, they were closest to modern newts but some could be several metres long; one species reached 5m. On the end of the primitive limbs, the number of fingers and toes had not yet standardised and different groups had different numbers, usually six to eight. Two of the best-known early fossils from this period are *Acanthostega* and *Ichthyostega*, although neither of these was our ancestor.

Acanthostega (meaning 'thorny covering') was only 7cm long and looked like an armoured newt. The head in particular was covered by thick bony plates like its fish ancestors. It had eight fingers on its front limbs and appears to have been completely aquatic as it still had a newt-like tail and internal gills like a fish.

Ichthyostega ('fishy covering') was a much larger animal, about a metre long with seven toes on its hind feet. It too had a finned tail like a newt and spent most of its time in the water, but *Ichthyostega* had no internal gills.

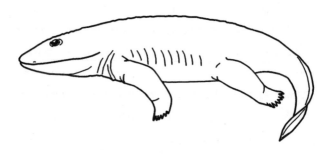

The early amphibian Ichthyostega

As the early amphibians spent more of their time on the land, their bodies changed. They still laid their eggs in water, and appear to have been aquatic when young, but in the adults the internal gills disappeared and the mid-line lost its fins as the adults became increasingly terrestrial. They still had scales on their undersides where they rested on the ground but the scales were lost from the rest of the body, as were the bony head plates. Without the support of the water and now weighed down by gravity, over the next 50 million years the limbs and the underlying bones of the shoulder and pelvis became stronger, as did the spine.

Early amphibians, held against the ground by their own weight, would have had difficulty lowering their jaw to open their mouth without first lifting the whole front of the body. Presumably because of this, a flexible neck region developed. This allowed the head to be lifted independently or to swing sideways. Fish can't do this. A sardine can't look over its shoulder.

Amphibians' ability to hear in air also improved, but no one knows what they were hearing or whether they were making sounds themselves.

Limbs

Many amphibians today (salamanders and newts) move on land in the same way as a fish, as do most living reptiles. They bend their body from side to side in such a way that a wave passes down the flank. In air, as once on a riverbed, the limbs act as points of contact with the ground and lever the body forward.

YOUR BODY

This fish-like movement on land leaves only two limbs on the ground at the same time, one at the front and the opposite limb at the back, but this is not very stable. To improve stability, the tail could be used as a third support to form a tripod, or the underside of the animal could be left resting on the ground. This would help but there would be too much drag on all but the smoothest surfaces, or when sliding downhill. Evolution solved this problem by changing the way some of our early ancestors walked. They now left three legs on the ground at the same time while raising the body. Modern salamanders and reptiles can use this method, although living amphibians may also use the two-point walk.

This change in the way of walking had profound consequences for the shape of our arms and legs. Instead of remaining just a peg-like projection, the limbs developed a bend, creating simple elbows and knees which curved the end downwards to engage more forcefully with the ground. The muscles also became stronger and allowed the end of the limb to push against the earth and increase forward movement. Our ancestors no longer relied only on the flexing of their spine and body muscles. The ends of the legs also changed. In earlier species, these tended to be straight continuations of the limbs. With the greater contact with the ground the ends now acquired their own bends and rudimentary wrists and ankles appeared.

Fingers and toes

About 360 million years ago, the world began to change and entered a time of land upheaval with the creation of enormous swamps. These swamps laid down today's vast coal reserves and led to this period of prehistory being called the Carboniferous ('carbon-bearing') age.

By this time, natural selection appears to have standardised the number of toes at five in most, if not all, amphibians. All subsequent four-legged animals would have five toes on each foot, although later groups often lost some. Curiously, today's amphibians have five toes on their rear feet but only four on the front and it is not known whether the front legs have lost a finger from an ancestral five or whether modern amphibians branched away early in evolution from an ancestral amphibian lineage – with more than five fingers – and developed their four fingers independently from the rest of us.

What we inherit from our amphibian ancestors

Our time as amphibians has left us with: a bendable neck; elbows; knees; wrists; ankles; five toes on each foot and five fingers on each hand (which led ultimately to the decimal system and percentages). This was also the time when we lost our fins, our gills and most of our fish scales.

Chapter Eight

When We Were Reptiles

Until now the ancient amphibians always had to return to water to breed. They all started their lives as frog-spawn. This changed for our amphibian ancestors about 310 million years ago when they evolved an egg with an outer membrane which prevented water evaporating but allowed oxygen to diffuse in from the surrounding air. These amphibians no longer had to return to the water to breed; they were becoming the first reptiles. A shelled egg like this was a major innovation, a space capsule for the embryo, carrying it into a dry alien environment in a bubble of the ancestral home.

These new reptiles (from the Latin *'reptilis'*,

meaning 'creeping') also developed a waterproof skin of light, flexible horny scales made mainly of keratin, unlike the heavy scales of their fish and amphibian ancestors. Keratin is the same substance that now forms our fingernails and hair, and birds' feathers.

These changes took time and it was about 280 million years ago before reptiles that were fully adapted to a life on dry land evolved.

The march of the reptiles

The problem for any animal on dry land is the force of gravity. A fish floats in water, effectively weightless. Today's four-legged animals either rest the weight of their bodies on the ground like lizards, with their legs sprawled on either side in what has been described as a permanent press-up, or they hold the body off the ground above vertical or near-vertical legs like cows or dogs. This move to a vertical leg began with some reptiles and would be perfected both by our reptile ancestors and by their cousins, the dinosaurs.

Our ancestor at this time appears to have been in a group of reptiles containing animals with large sails on their backs, the pelycosaurs (Greek for 'basin lizards').

WHEN WE WERE FISH

Dimetrodon, *a pelycosaur 3m long*

Some biologists have suggested the sails may have been a type of solar panel to maximise the collection of heat from the sun in these cold-blooded animals (but not all pelycosaurs had such sails; some later species managed without them). Pelycosaurs were large reptiles – some could reach 3m in length – and were the dominant land animals of their time, although most seem to have been predators eating fish and amphibians which suggests they still lived near water.

By 240 million years ago, pelycosaurs had all but disappeared, but one group had survived and had changed sufficiently to be given its own complicated scientific name, the therapsids (from the Greek, 'arched beast'). We are descended from the therapsids.

YOUR BODY
Limbs and spine

Our time as therapsids significantly shaped our spine and ribs and brought further important changes to our arms and legs. Therapsids had a less sprawling posture than their predecessors, with their legs held further under their body. They appear to have spent more time away from water than the pelycosaurs and, as their ability to move over the land began to improve, several things happened that changed our bodies forever. The first was the removal of parts of the skeleton that would prevent faster, more energetic walking. Reptiles had inherited a side-to-side flexing of the spine from the amphibians and they still had ribs on most of the vertebrae between the front and hind legs. As the back legs became stronger and played a more active role in pushing the body forward, there would have been a tendency for the ribs in front of each hind leg to be crowded together as the backbone bent sideways. Natural selection solved this crowding by removing the ribs in this part of the spine. This created a recognisably different region of the backbone, which we now call the lumbar region, and left a rib cage only behind the front legs. This solution to more efficient movement is why today we have no ribs in our lower back, or across our stomachs.

WHEN WE WERE FISH

Skeletons of a pelycosaur and a therapsid, showing the loss of the ribs in front of the hind legs

The second change that occurred as the therapsids became better walkers was that the feet moved to a position under the body. An animal that rests its whole body on the ground cannot move very quickly or efficiently, and lifting the body off the ground on legs that project sideways needs large muscles and a lot of energy. Try it yourself. Lie face down on the ground then lift your body in a press-up. Success is easier if you put your hands on the ground near each shoulder. The further you reach to the side before placing your hand, the harder it becomes to lift yourself and hold that position. An animal that holds its legs vertically or nearly vertically under its body can use the bones like a column to hold its weight while the muscles are used only to make small adjustments of posture and balance. We can stand at attention for a long time, but squatting with our thighs parallel to the ground soon becomes very tiring.

Elbows and knees

In our therapsid ancestors the repositioning of the limbs was accomplished by rotating the front legs backwards, so the upper arm pointed back along the flank, and by rotating the hind legs forwards so the thigh pointed forwards along the flank.

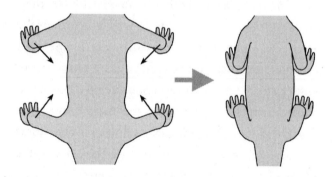

The different directions of rotation for the front and back legs left the joint in the middle of the front leg bending in the opposite direction to the joint in the middle of the hind leg. This is why our elbows and knees still bend in opposite directions today, as do those of all mammals and the reptiles' other descendants, the birds. The legs of birds and two-legged dinosaurs like *T. rex* look as though they have knees that bend backwards rather than forwards, as do the hind legs of many mammals (cats, dogs, horses), but they have not. Why they have not will be explored later.

About 240 million years ago, a new group of therapsids appeared, the cynodonts (pronounced 'sign-oh-donts', Greek for 'dog-toothed'). Some were herbivores but most were carnivorous, hence their dog-like teeth. The cynodonts were to be the dominant land carnivores for the next 10 million years. Their limbs were held directly under their bodies and there is some indication they were on the way to becoming warm-blooded. Other reptiles, like all vertebrates before them, relied on the sun to warm their bodies and raise their metabolic rates. Cynodonts also had a better ear for hearing sounds in air, with an early eardrum. Initially, many cynodonts were the size of a large dog but later species became smaller, many no more than 25cm long.

The therapsids were now the dominant land animals, having overtaken other reptiles and the surviving amphibians. However, by about 210 million years ago, their reign was over and most had already gone. The last would become extinct within another 60 million years, but not before one group of cynodonts had evolved into a new type of small rodent-like animal. They had become the earliest mammals. One day, their descendants *Homo sapiens* would inherit the Earth, but, before that, from another branch of the reptile family, a new king had arrived. This was now the Jurassic, and the Jurassic was the age of the dinosaurs.

What we inherit from our reptile ancestors

Our time as reptiles has left us with: a waterproof skin with no scales; a lumbar region of our spine with no ribs; elbows and knees that bend in opposite directions; an eardrum; and the first stages of becoming warm-blooded. It was also when we lost the need to travel to open water in order to have children.

Chapter Nine

Mammals

Our early mammal ancestors lived alongside the dinosaurs for 150 million years without making their presence felt. Only when most dinosaurs died out about 65 million years ago did mammals step out of the shadows and begin to fill the vacuum. They would soon develop into the diverse range of forms that dominate the land in most parts of the world today.

Mammals are characterised by having, among other features, hair and the milk-producing mammary glands after which they are named ('mamma' is ancient Greek for 'breast'). Today, there are three groups of mammals: the monotremes; the marsupials; and the placental mammals. The least common of these are the monotremes (Greek for 'one-holed' because their intestine, bladder and reproductive tract all exit through the same opening). Their only living representatives are the duck-billed platypus and two species of spiny anteater in Australasia, which lay eggs but suckle the young with milk when they hatch.

Marsupials give birth to tiny young at a very early stage of development and nurture them in a pouch (a marsupium) where they continue their development nourished with milk from mammary glands.

Placental mammals such as ourselves keep the young inside the body until a much later stage of development, nourished through a placenta.

None of these three groups is more advanced than the others; each is one solution to mammalian reproduction, although the earliest mammals probably laid soft-shelled eggs like the monotremes, as did their reptile ancestors.

Most mammals may have stopped laying eggs and kept them inside the body during development because there was an advantage in being able to move freely while breeding rather than being confined to a nest. Perhaps a nomadic lifestyle or the ability to breed in the treetops, away from the dangers on the ground, gave most early mammals a better chance of survival. Whatever the reason, the shell of these internal eggs disappeared and other modifications evolved.

Mammary glands

Mammary glands illustrate another feature common in mammals, the presence of glands in the skin. Reptiles and birds have few skin glands but mammals have many, of various types. Historically, breasts are simply concentrations of enlarged skin glands. Milk is modified sweat.

The number and distribution of breasts among mammals is very varied. Humans have two, not four or six or eight or more like other mammals (some opossums have more than 20). Breasts are always on the body's under-surface but in some mammals they are along the full length (pigs, dogs), in some they are only between the hind legs (cows, horses, sheep), while in humans and other primates they are only between the front legs.

Hair

Hair was a new development for the mammals but its exact origin is obscure. The other offspring of the reptiles, the birds, developed feathers which are almost certainly modified scales. Birds also kept unmodified scales, as is obvious from a chicken's legs. Some mammals have skin scales, such as those on rats' tails, but whether hairs are modified scales is not clear.

However it evolved, hair provided both insulation in the newly warm-blooded mammals and colour for protection and signalling. Today, there are: black coats (panthers – which are just a black form of the leopard or jaguar); near white (polar bears and other arctic mammals in winter); black and white (zebras; skunks; giant pandas); grey (wolves); and many shades of brown, although some verge towards exotic yellows and oranges (giraffes, tigers and spotted cats). All these colours are produced by one pigment, melanin, which has two forms: one is black or shades of brown; the other is red to yellow. (Incidentally, melanin is also the pigment that colours human skin.) No hair is ever green, but most mammals are red-green colour-blind anyway. They see blue and yellow but have difficulty distinguishing red, green, orange and brown. Among mammals, only primates have colour vision like ours. To a fox, a rabbit is the same colour as grass. To a rabbit, so is a fox.

Humans are unusual, when the European racial type is considered, in that hair colour seems to follow the full range of mammal possibilities but within one species, although white hair or the presence of white or grey streaks in otherwise dark hair are usually signs of age, and each of us has hair of one colour, not striped, spotted or piebald. The only other mammals that show such a wide range of hair colour within one species are domesticated breeds which we have deliberately manipulated to produce the colours and patterns.

MAMMALS

Various reasons have been proposed for some wild mammals having striped or spotted coats. The usual reason given is camouflage, although why cheetahs have spotted coats while lions, in the same grassland environment, have plain coats is not obvious unless the fact that lions (or correctly lionesses) hunt as a team is relevant. It also seems strange that all species of spotted cats have different patterning. If the patterning was for camouflage we would expect different species to have evolved the same pattern. It may be that, for a colour-blind mammal, using contrasting patterns of dark spots, rings or stripes is their way of signalling their identity to others of their own species. Other vertebrates can do this with colour.

Limbs and the spine

Early amphibians, with limbs that projected from the sides of the body at right angles, could only walk by moving their arms and legs in an overarm stroke similar to the arm movements of someone swimming freestyle (a style also known appropriately as 'the crawl'). This needs a large circular movement away from the side of the body.

When our early mammal relatives (or reptiles becoming early mammals) turned their limbs downwards under the body, their stepping became a flatter arc in the forward–backward plane, fully in the direction of travel.

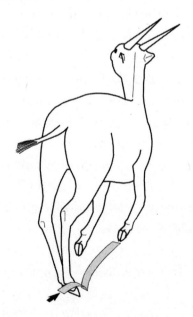

MAMMALS

With the legs underneath the animal, its underside was no longer dragging along the ground and its body no longer lurched from side to side with each step. The spine stayed in a straight line forwards and backwards and the limbs swung straight forward, with the muscles of the legs now providing the thrust. Despite this extra duty, the leg muscles actually became smaller. This could happen because the weight was now supported on the near-vertical bones. The muscles were now for movement, not support. These changes made walking extremely efficient, needing much less energy and effort than the ancestral crawl.

This new posture became almost universal in the early mammals and most mammals today hold the body well clear of the ground on near-vertical legs. Some reptiles also adopted this upright posture, the dinosaurs being an obvious example. Their descendants the birds also show this trait. Most surviving reptiles retain the ancestral sprawl but some can lift themselves on their legs for rapid bursts of speed. Crocodiles sprawl when they are lying on a riverbank with their underside resting on the ground, or when they are moving at a slow walk, but when they need to run they swing their legs to the vertical and lift their bodies. This not only reduces drag, but also increases stride length. Crocodiles may look lethargic on land but it's best not to test their racing credentials.

YOUR BODY
Why we can shrug our shoulders

Towards the rear of a mammal's skeleton, the pelvis is welded directly to the spine by fused bone. This makes it very stable for the attachment of the now strong back legs. By having the pelvis attached directly in this way there are no soft tissues to absorb any thrust, so all the push from the back legs propels the spine, and hence the body, forwards. The front legs are not as important for providing forward force; they are used more for changing the animal's direction. For this they need mobility. Mammals have their front legs attached to the bones of the shoulder, especially the shoulder blades, but unlike the pelvis these are connected to the spine by muscles. This difference is easily demonstrated by the fact that we – as mammals – can shrug our shoulders whereas we can't shrug our hips. In four-legged mammals, the connecting muscles of the shoulders also act as shock absorbers when the front legs hit the ground at speed. This reduces the jarring to the skull and eyes, which need to be fully operative when sprinting. This shock-absorbent arrangement also has advantages for us. If your shoulder was fused directly to your spine like your pelvis, using a pneumatic drill or a jack-hammer would literally scramble your brains.

A pelvis that is fused to the spine did create one problem for early mammals. As the hind leg of each

MAMMALS

side was lifted to take a step, the whole pelvis had to be tilted to give the necessary ground clearance to the foot. Meanwhile, at the other end of the body, the diagonally opposite front leg was just completing its movement and the shoulder at that side was still raised. As a result, walking forced the spine to undergo alternate twisting along its whole length, the rear twisting one way, while the front twisted the other. This action is like wringing the water from a wet cloth. It is thanks to this requirement that we can now turn our hips in one direction while turning our shoulders in the other. Without this ability, we could never play golf. Other vertebrates cannot do this. Similarly, only mammals, with our twistable spines and our repositioned legs, can lie down on our sides (and get up again). Reptiles can only lie on their stomachs.

With the limbs now under the raised body and moving backwards and forwards, another change occurred in the spine. Instead of bending from side to side like a fish, which would no longer help walking, the spine began to flex up and down. When the rear end of the spine flexed downwards at the same time as a rear leg was stepped forwards, the foot met the ground further forward than if the spine was rigid and only the leg was moving. This increased the length of a stride making walking and running even more efficient. This new vertical bending ability is also the reason why we can now lean down and touch our toes.

This move to a vertical flexing of the spine later had profound consequences for one particular group of mammals. When the four-legged ancestor of the whales and dolphins returned to the sea and once again used its tail for swimming, this tail now swept up and down, not side to side like its fish ancestors.

Why we ride horses not cats

Few mammals have perfected this vertical flexing more than the cheetah. When it sprints, its spine bends like a bow, first upwards then downwards. As the middle region of the spine bows downwards, the front legs are stretched far out in front of the animal, maximising its reach. With the front feet planted on the ground the spine curves in the opposite direction, bowing upwards at the same time as the hind feet are pushed forwards.

Thanks to this spinal flexibility, the rear feet actually touch the ground ahead of the front feet. The muscles of the hind legs can then pull the animal forward, helped by the muscles of the back as the spine springs straight and continues through into the opposite curve again. This is similar to the actions of an olympic rower, reaching forwards until their hands virtually touch their feet, then arching their back and pushing with their strong (hind) leg muscles.

This spinal flexibility contributes to the cheetah's great speed, but we should be grateful horses use a different system. If the spine of a horse moved like the spine of a cheetah, it would be like trying to ride an ejector seat.

Horses and other hoofed mammals hold their spines relatively horizontal when running. Unlike the cheetah, they are not built for rapid sprinting; they have evolved for sustained travel over open country at a steady run. A cheetah can race faster than any other animal, more than 110kph (68mph), but only for short bursts. A horse can maintain a steady canter for hours. In the horse several modifications of its mammal body allow this.

First, a horse's legs have become longer as the foot has stretched and the heel has lifted off the ground. Many mammals permanently walk on their toes like this, including cats and dogs, but the horse has taken this one stage further. The horse has pointed its toes until it stands on the very tip, like a ballerina.

YOUR BODY

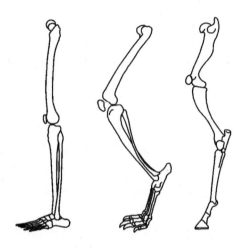

Hind legs of a human, a dog and a horse (not to scale) showing our flat-footed stance with the heel on the ground, the permanent tip-toe stance of dogs (and cats) and the ballerina stance of horses

This posture, together with a great elongation of the bones, increases the stride length and produces energy-efficient speed. In the hind leg, what appears to be the animal's knee, pointing backwards, is actually its heel, now permanently off the ground and halfway up the leg. The knee is at the top of the leg next to the body, and points forward as expected. In the horse's front leg, what appears to be a knee is actually a wrist. The elbow, like ours, points backwards and is also now at the top of the leg.

Second, the horse's leg has been made lighter by a reduction in the number of bones. The side toes on all four feet have shrunk almost to nothing and only the middle toe remains. The number of bones in the foot has been reduced and the two parallel bones of the lower leg have become one. The weight loss is important because every time the horse takes a step, especially when running, it must lift its leg and throw it forward. The heavier the leg is, especially near the foot, the more effort the horse has to make to get it moving. A lighter leg can be lifted and moved quickly with much less exertion.

Third, but for the same reason, the heavy powerful muscles which operate the leg are not near the bones they need to move. Our calf muscles can be very heavily developed and are at the end of our legs, where they must be lifted every time we take a step. In the horse, the massive muscles are all at the top of the leg in the powerful hindquarters and shoulders. These muscles are connected to the bones of the lower legs and feet by light, wire-like tendons (also sometimes called sinews). The muscles pull on the tendons which in turn pull on the bones and move the legs. These adaptations together give the horse its light and slender limbs which can be stepped out at high speed for long periods without unduly tiring their owner.

Sometimes tendons can assist in movement directly because they have a natural elasticity. The tendon we are most familiar with is our Achilles tendon at the back of the heel. In the kangaroo, the Achilles tendon is very long and its elastic properties are used when the animal is hopping. Much of the bounce comes from the natural recoil of the tendon, not from energetic contractions of the muscles. This allows the kangaroo to move quickly using very little energy, like a person on a pogo-stick letting the spring do all the work.

Speed and toe loss

Modern horses stand on the tip of the third toe of their fore and hind feet. Their ancestors had already acquired this one-toe stance about 5 million years ago after leaving their forest origins and adopting a life on the open grasslands.

Many modern animals, especially the hoofed mammals, show a reduction in the number of toes from the ancestral five. Rhinoceroses walk on three toes, while cows and deer walk on the tips of two toes which look like one hoof with a cleft in the middle and give them the name 'cloven hoofed'. Toe reduction is not restricted to mammals. *Tyrannosaurus rex* and many of his relatives walked on three toes, and among the birds the ground-running ostrich now has only two toes.

MAMMALS

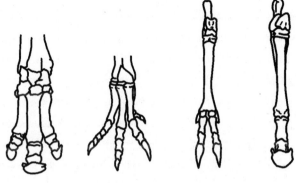

Toe reduction in a rhinoceros, the carnivorous dinosaur Allosaurus, a deer and a horse (not to scale)]

We humans did not pass through an evolutionary stage requiring us to move swiftly over open country. For most of our history we have been creatures of the undergrowth and trees. As a result we have kept all our toes and fingers and still walk on the soles of our feet with our heels touching the ground. Few other mammals still do this, although bears do too.

Warm blood

As mammals, we are warm-blooded animals, although what we call 'warm-blooded' animals are actually animals that can maintain a constant internal temperature regardless of the temperature of the environment. Both mammals and birds do this and should really be called constant-temperature animals.

Similarly, fish, amphibians and reptiles are not cold-blooded, they simply rely on their surroundings or on the direct effect of sunlight to provide the warmth they need for their metabolism. If the environment is cold they tend to be sluggish, but if it's warm they can be very active. Anyone who has seen a tortoise dragging itself laboriously across an English garden then travelled to the equator and seen tortoises sprinting past each other under a tropical sun does not need this explaining further. Keeping tortoises as pets in Britain is now discouraged, doubtless to the great benefit of the tortoises.

The advantage of being warm-blooded is that an animal can be active regardless of the surrounding conditions. This can be especially helpful if it allows them to exploit the food that is available when the sun is not shining. Perhaps mammals acquired their warm-blooded nature to exploit this gap in the market when reptiles were sluggish in the cool night air.

The disadvantage of being warm-blooded is that an animal must burn a lot of energy just to stay warm. It therefore needs to eat at regular intervals even when it is not active. A snake may need a meal only once a month but most mammals and birds will die within days without food.

A constant-temperature animal like a mammal must also be able to keep its temperature constant even when it is working hard (running or climbing) or when the environment is too hot. Many mammals have solved this problem with sweat glands in their skin. The moisture these exude evaporates and carries away body heat. Heat is also lost by evaporation from the lungs, mouth and nose, which is why many mammals also pant when they are hot. Different species of mammal rely to different extents on perspiring and panting. Dogs only have sweat glands on their feet so they pant to lose heat, mainly through their nose and lungs. They also allow their long tongue to hang out of their mouth to add to the evaporation. We also pant when we are hot but we rely much more on perspiring and fanning ourselves to evaporate the sweat more quickly (which is why a fan feels cool). Of course, we can also shed our outer layers of clothing, a technique not available to other mammals.

Male gonads

At the risk of being indelicate, one curious aspect of the mammal body is the descent of the testes to hang outside the body cavity in a bag of skin, the scrotum. This happens in most species, usually permanently but sometimes – as in squirrels and some bats – only during the breeding season. No other organs do this in mammals. We don't have two purses hanging out of our backs containing our kidneys, or a large pendulous liver swinging from the bottom of our rib cage. Even women's gonads (their ovaries) don't project from the body; so why men's gonads do is rather a mystery.

The usual explanation is that the testes must be kept cool because sperm formation is most successful at a lower temperature than that inside the body (1–3°C lower), but elephants, armadillos, sloths, whales, seals and sea lions all have internal testes. Birds also have internal testes and they have a higher body temperature than mammals. Chickens and budgerigars maintain a temperature of 41°C compared to a human's 37°C. Surely, if testes evolved to be inside the body, they should have evolved to work at the temperature inside the body. 'To stay cool' seems to be a poor explanation for their eviction. It might be more logical to conclude that sperm formation is most efficient at a temperature lower than the body's core because the testes in mammals have evolved to work outside the body at that lower temperature, not the other way round. However, if that is wrong and the problem for sperm formation *was* the increasing temperature of the body in some reptiles, perhaps it's just the turn of men to suffer the indignity of an evolutionary solution that worked, rather than a perfect solution. As body temperature rose in pre-mammal and pre-bird reptiles, perhaps the birds solved the problem by changing the physiology of sperm formation in the testes whereas the mammals solved the problem by pushing the testes out of the body. This may be embarrassingly unattractive and most inconvenient, but it worked so it survived.

Whatever the reason for the descent of mammals' testes, it has only been possible because mammals rotated their legs towards their mid-lines and substantially raised their bodies. Amphibians and reptiles seldom have any ground clearance.

What we inherit from our four-legged mammal ancestors

Our time as four-legged mammals has left us with: warm blood; hair; perspiration; breasts; pendulous testicles; and the ability to rotate our shoulders and hips in opposite directions and to touch our toes. We also develop in our mother's womb and no longer hatch from an egg in a nest, and we drink milk as newborn infants.

Chapter Ten

Mammals

We are primates. The earliest origins of this roup of mammals have still not been uncovered in the fossil record but, by 55 million years ago, there were small squirrel-like primates living in North America and Europe, possibly related to rodents.

We should not be surprised the fossil record is poor for primates. They seem to have lived mainly in woods, and woodland floors are not favourable sites for turning bones into fossils. The soils tend to be mildly acidic, which can attack bone, and plant roots and permanent damp aid decay rather than preservation. If the climate changes and the trees disappear, the remaining soils can erode quickly, destroying any bones that have survived. Fossils are more likely to be found in sites where layers of sediment or ash were deposited on top of any remains and lack of oxygen prevented decay. In wooded country this is not likely to happen.

With the exception of ourselves, most modern primates are still tree or forest dwellers. The two groups we are most familiar with are the monkeys and the apes. Monkeys usually have a long tail while the apes have lost theirs. This absence of a tail is such an obvious feature that one species of monkey that has also lost its tail is known falsely as the Barbary Ape. Monkeys use their tail for balance when walking along the tops of branches and in some it is prehensile and can be wrapped around boughs to act as an extra arm. Some species can even hang solely from their tail while collecting food. Apes, being much larger, tend not to walk along branches. They usually swing under the branches from their arms.

In most mammals, the limbs hold the body off the ground and are compressed along their length, but in a primate hanging under a branch the arm is being stretched by the full weight of its owner. To prevent this damaging the muscles of the shoulder, which did not evolve to take such tension, primates have developed a prominent collarbone (or 'clavicle') which helps to spread the load evenly. Running mammals such as horses and antelopes – which never swing from branches – have lost their collarbones completely.

Unlike other mammals, which are red-green colour-blind, all primates have colour vision like our own. This allows these tree-dwelling fruit and flower eaters to recognise ripe red fruit and red flowers against a background of green foliage. Primates have come to rely much more on sight than on their other senses and our sense of smell is not particularly good, which is why we use trained bloodhounds and other dogs when scent is important to us.

Monkeys are found today throughout the warmer parts of South America, Africa and Asia but apes are found only in Southeast Asia (the gibbons and the orang-utans) and central Africa (the chimpanzees and the gorilla). The orang-utans, chimpanzees and gorilla are together known as the 'great apes'. There are two species of orang-utan, two species of chimpanzee – the Common Chimpanzee and the Pygmy Chimpanzee or Bonobo – and one species of gorilla, although the gorilla has three sub-species – the Western Lowland Gorilla, the Eastern Lowland Gorilla and the Mountain Gorilla.

Of these apes our closest living relatives are the chimpanzees with whom we share about 99% of our DNA. Put another way, we are only 1% not chimpanzees. However, there is more to an animal's appearance than just the sequence of its genes. We differ in appearance from a chimpanzee by much more than 1%. Appearance depends not just on the sequence of the genes but on how the instructions from those genes are put into effect. Identical twins have identical genes and look identical. A person and a chimpanzee have almost identical genes yet look quite different. Presumably in identical twins the way the genes' identical instructions are put into effect is also identical, whereas in humans and chimpanzees it is the way the almost identical instructions are put into effect that is different. Evolution in this case does not appear to have been the result of natural selection editing the genes but of selection editing how they work.

Humans and chimpanzees evolved from the same ancestral species only a few million years ago and chimpanzees have consequently attracted a lot of attention from scientists interested in our own origins, especially those studying language. Chimpanzees' vocal cords are not sufficiently like ours to allow them to speak, even if they wanted to, but some scientists claim to have taught chimpanzees to communicate with sign language. These chimpanzees have not mastered grammar or sentences but give the signs for objects and some other words. One, called Washoe, can form 240 different signs and sometimes puts them together to make new combinations she has not learned, such as referring to a watermelon as a 'Drink Fruit'. In the wild, chimpanzees communicate with hoots and howls, mainly conveying an emotional state, but they also have a subtle understanding of facial expressions, body language and gestures. Some will even invent their own signals. This is presumably why they can adopt human sign language in captivity.

Like us, and unlike the gorilla, the two species of chimpanzee supplement a diet of leaves, fruit and seeds with up to 10% meat. Bonobos will eat rodents and snakes but Common Chimpanzees will collaborate to hunt and kill monkeys, pigs or antelopes.

The similarities between ourselves and chimpanzees in social organisation, hunting collaboration and communication skills suggest we acquired many of our social characteristics when we lived in the forests, not when we became *Homo sapiens*, and support the view that we are not uniquely distinct from other species but are merely one part of a continuous gradient of types.

The primate hand

In most non-primate mammals, the only grasping organ is a pair of jaws lined with teeth, or the two front paws pushed against each another. It is only the primates that have a highly developed opposable thumb. The primate hand is therefore an integrated grasping structure in which the fingers work as a team. Here is a demonstration, but first read this warning:

WHEN ATTEMPTING THE FOLLOWING, *DO NOT* HOLD THE MIDDLE FINGER DOWN WITH YOUR OTHER HAND WHILE BENDING THE INDEX FINGER. THIS WILL DAMAGE A TENDON IN YOUR FOREARM.

Having read the warning, lay one of your hands flat on its back on a flat surface. Now bend the little finger to touch the palm. Note how the finger next to the little finger has also lifted, you cannot prevent it. Similarly curling the index finger to touch the palm raises the middle finger. This happens because the hand has evolved to grasp, and this can be achieved with minimum effort and maximum speed if the fingers are connected to the same closing mechanism. In our hand, the closure is led by the little finger. If you fold all your fingers one at a time until their tips touch your palm in a fast continuous grasp, it is much easier, and feels much more natural, to start with the little finger and finish with the index finger than vice versa.

Opposing this team of fingers in primates is our

opposable thumb. Opposable digits occur again and again in the animal kingdom but few groups have this facility in all their members. Perching birds have opposable toes, but some species have three toes opposed by a fourth and others have two opposed by two others. Some reptiles, such as the branch-walking chameleon, also have opposable digits. Among the invertebrates, opposable grasping organs take a variety of forms, the most obvious being the claws of crabs and scorpions, and the front legs of insects such as the praying mantis. All these structures are used to manipulate objects ('manipulate' from the Latin '*manus*' meaning 'hand').

We only have an opposable digit on our front legs; other primates have 'thumbs' on their front *and* rear legs. We lost the opposable toe on our feet when we left the trees but the large size of our big toe indicates its past special role.

Humans have the most dexterous hands of any ape. We can easily touch the tip of our thumb to the tip of the other fingers because our thumb is relatively long. Chimpanzees have a much shorter thumb which can manipulate objects but not as precisely. The shorter thumb arose because when apes swing under branches they do not usually grip the branch. Instead, they simply make a hook of the bent fingers and loop this over the bough. Their short thumb, near the wrist, does not then interfere with this hook. Chimpanzees only grip a branch when they are moving slowly or standing on top of it, but, even then, like most great apes, they do not grasp but commonly lean on their knuckles, as they do when walking on the ground.

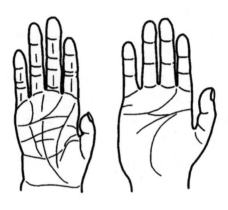

Hands of a chimpanzee and a human

The primate hand has one other adaptation for manipulation. In most species, the claw has been changed into a flat nail. This protects the end of the finger but allows the sensitive ball of the tip to feel and control even hard polished surfaces, relying on friction for a sound grip rather than on spiking the item with a needle-like point. To increase the friction, the skin in this region is finely corrugated, giving us our fingerprints.

PRIMATES
Arm mobility

The grasping hand evolved because it allowed us to move easily through the trees, but trees are not very predictable environments. Branches can grow in any direction in a chaotic tangle of handholds. To meet the needs of a primate's life, natural selection had to find a way of getting a primate's grasping hand into virtually any position. It achieved this by modifying the whole arm and giving it an astonishing range of movement.

Primate arms can rotate in all directions. Nothing demonstrates this better than watching the arms of swimmers in an individual medley. Butterfly, backstroke, breaststroke and freestyle together give a graphic demonstration of virtually the full range of primate arm movement. The only stroke missing is the one forced upon most other mammals – the 'doggie paddle'.

When we are not swimming, we can put one hand on the opposite shoulder then swing our arm horizontally through three-quarters of a circle to point towards the rear. We can point the arm towards the ground then swing it in a full circle across the front of our bodies (if we lean our heads back out of the way) or in a full circle next to our side. For the few positions a primate can't reach using only the mobility of its arms, the familiar twistability of the mammalian spine, highly developed in primates, comes to the rescue. But the story did not end with an arm that can reach in virtually all directions. Branches may not only be in any direction from our body, they may also be growing in any orientation: horizontal, vertical, diagonal, leaning away, leaning towards us. To cope with this, the hand of the primate must be able to twist into any position before it grasps. Natural selection solved this problem by easing the connections between the two bones of the forearm, the radius and the ulna. These bones could then move to cross each other and the forearm could twist along its length. This is why we can now hold our hands palm up or palm down. Other mammals cannot do this. When our palms are up, the two bones are parallel; when we rotate our thumbs inwards to turn our palms down, the two bones cross. If you grip your forearm with the other hand, you can feel them doing this, especially near the wrist. When this rotation is added to the flexibility in the other joints of the arm and shoulder, it allows us to hold our hand straight out in front of our bodies and twist it through more than 360°.

Even that was not the end of the primate arm's evolution. We can also wave our hand from side to side using only the wrist. Being able to move the hand like this allows us to change our grip from a vertical branch to a horizontal branch running away from our body without moving our forearm at all.

PRIMATES

As if all that wasn't enough, we can move each arm independently of the other in almost any combination, one reaching up and forwards while the other reaches down and backwards or to the side, or both arms reaching to the same side at different levels, and our ancestors could reach out and grab other branches with their grasping feet at the same time.

This incredible range of movement is a remarkable achievement not found in other animals. Horses and dogs can only move their front legs through a small swing in a forward–backward direction and even the gymnastic squirrels are severely restricted compared to people.

Seeing in 3D

For any animal living in trees and having to jump from branch to branch, it is imperative not only to know where the next branch is, but how far away it is. For this sort of depth perception, an animal needs to see the same image through both eyes. Only when two eyes see the same object from slightly different viewpoints (the distance between the eyes) can the brain judge distance. For this reason the eyes of the primates moved to the front of the face, side by side, where they remain in us today. Other types of mammal have eyes suited to their own lifestyles. Many herbivores liable to attack by predators (for example, rabbits) have eyes set high on the sides of the head, pointing sideways. This gives them a 360° field of view, but very little depth perception.

Primates have maximised their depth perception at the expense of such wide fields of view. To compensate for this, they have continued to develop a flexible neck and now have one of the most mobile heads in the animal kingdom. Few mammals can tilt their head backwards as far as a primate. We still cannot see directly to the rear but being social animals most primates can rely on others warning them of dangers approaching from behind.

Standing upright

Tree-dwelling primates stand upright. This is not surprising in an animal that has to reach for branches in a treetop existence. They stand on one branch and reach up for a higher hand-hold. This ability can be seen on the ground too where most monkeys and apes can stand, or even walk, on their hind legs, although this is not their usual method of moving and usually looks clumsy and inefficient. Nevertheless, when we think of our ape-like ancestors leaving the trees to walk upright, we should not think of this as a new posture. It was an extension of a posture already found in other primates. Our ancestors merely switched to using this on the ground more than on the branches. Just as the fish already had lungs and limbs before they left the water, so we could already stand upright before we left the trees.

Male genitalia

At the risk of being indelicate again, male primates have one more unique feature – a permanently exposed penis. This may be related to the ability of primates to stand on their hind legs, which naturally displays their under-surface. In reptiles, the penis is usually fully inside the body unless in use. Other mammals (stallions, bulls) may have an indication of the location of the penis, but the organ itself is usually out of sight inside the body; few mammals have it permanently exposed.

What we inherit from our tree-dwelling primate ancestors

From our tree-dwelling ancestors we have inherited: grasping hands with an opposable thumb and great manual dexterity; fingernails and fingerprints; an extremely wide range of arm movement including a rotatable forearm; a prominent collarbone; a large big toe; an ability to balance on our hind legs; a very mobile neck; forward-facing eyes and depth perception; the ability to distinguish red from green; and permanently displayed male genitals. This was also the time when we lost our ancestral tail.

Chapter Eleven

'Hominids'

We are thought to have evolved from our ape-like ancestors in Africa. This was originally suggested because most living great ape species live in central Africa; only orang-utans live on another continent. Modern analyses, including DNA studies, support this view of an African origin.

There appear to have been many ape-like and potentially human-like primates living in Africa during the last 10 million years. We call those that we classify in the same family as ourselves 'hominids' because the scientific name for our family is *Hominidae*. Until recently, we had a rather high opinion of our own uniqueness and from the species alive today we chose to include only ourselves in this family, although we were prepared to tolerate fossil species if it pleased us. Nowadays, some people take a rather more realistic, and humble, approach to classification and are prepared to recognise the great apes – the chimpanzees, the gorilla and the orangutans – as living hominids.

Recent analysis of fossils and DNA suggests our ancestors separated from the ancestors of the chimpanzees about 5–7 million years ago but there is frustratingly little fossil evidence for the period from 10 million to 4 million years ago (because these were woodland animals). What scientists can say about our ancestors is that evolution modified the architecture of their ape bodies in several ways: the jaw became less prominent with smaller canines and more emphasis on large chewing molar teeth; the brain became larger; and of course they became increasingly upright and two-legged and lost the grasping foot of other apes.

Their dense covering of body hair also vanished. We still have hair all over our bodies but in most people it is so short and fine it is virtually invisible. Even in hairier individuals, it is too sparse to act as insulation and we need clothes in all but the hottest climates. Anthropologists do not yet know when in our evolutionary history we lost our fur, or why, because hair does not fossilise. Whether it was related

'HOMINIDS'

to moving out of the cool shade of trees onto the open plains is pure speculation.

There are currently two theories regarding our evolution from ancestral apes. The first is that the characteristics of the human body evolved progressively from an ape-like body only once. This is called the linear (or tidy) model of human evolution and is probably how most of us would think of it happening. The second, bushy (or untidy), model suggests that features we tend to think of as human – walking on two legs, a large brain, small jaws resulting in a flat face, loss of hair, great manual dexterity – may each have evolved more than once in different groups of apes, only one of which survived to become our ancestor. This second view would suggest that a fossil of a primate that walked upright and had a flat face need not necessarily be directly related to us. It might have been a species from a different ape line that later became extinct. The untidy model would also predict there could be fossils of various types of primate, from the same period, showing different combinations of human-like characters: some with large brains but walking like a chimpanzee on all fours; some with small brains but walking upright; some with large brains but large jaws and canine teeth like an ape; and other variations.

At present, there are not enough fossils from the critical period of 10 million to 4 million years ago to test which of these two theories is more likely. The past decade has seen an increase in the number of new fossils from this hominid 'dark age', most from East Africa, but virtually all the fossils discussed in books on human evolution are only from the past 4 million years. From these fossils we do know that our ancestors were walking on two legs by at least 3,750,000 years ago. Our own species, named *Homo sapiens* by the 18th-century Swedish naturalist Carl Linné ('Linnaeus'), first appeared only about 100,000–200,000 years ago as a descendant of an earlier hominid, *Homo erectus*, or 'upright man'. *Homo sapiens* is Latin for 'wise man' – a rather conceited name.

The human cradle

It is not yet clear where in Africa our ancestors evolved. It is often said this may have been the Rift Valley region of East Africa (in the area covered by modern Tanzania, Kenya and Ethiopia) but we must be careful with this interpretation. We know human and pre-human fossils from many eras exist there, but that does not mean that was the evolutionary cradle for our species. The international community has tended to focus its attention there in a very

'HOMINIDS'

reasonable attempt to obtain as much knowledge as possible from an area where information is known to exist. Anthropologists have not conducted the same detailed searches in all other parts of Africa.

Out of Africa

Moving forward in time to *Homo sapiens*, we know rather more about the history of our own species, especially since we have been able to compare DNA from different populations worldwide.

Some scientists now believe every person on the planet other than those native to Africa is descended from a relatively small band of humans that crossed the mouth of the Red Sea into the southern Arabian peninsula about 60,000 years ago. Within 5,000 years, their descendants had reached China and Southeast Asia and in less than another 10,000 years they were in Australia. By almost 50,000 years ago, some had moved through the Middle East to colonise Europe and others were spreading across central Asia. By a little over 20,000 years ago, they had reached North America and by 15,000 years ago were as far south as southern Chile. Every race of humans we see today outside Africa has been produced by natural selection from that one ancestral group, and that group was African. Little more than 60,000 years ago, every person on the planet was dark-skinned, dark-haired and dark-eyed.

YOUR BODY

Even today the natural human condition is to have dark hair and dark eyes. Only in a small area covering northern Europe, Scandinavia and Russia west of the Urals did humans evolve with hair and eyes that were not dark. This relatively small geographical region is also where the skin is palest. The weakening of the pigmentation of this aberrant racial group seems to have been a very localised development, although emigration has now spread these characteristics to many other parts of the world.

Sixty thousand years may not seem a very long time to evolve Arab, Mediterranean, Asian, Oriental, aboriginal Australian, native American and other (especially the unusual European) features, but it has been achieved because the differences between all of us are literally just skin deep. Despite our racial distinctions, we are still very much one species.

Chapter Twelve

Your Body Today

Now we have considered how our bodies arrived at their present shape, we can consider that shape in a little more detail, starting with the familiar backbone.

Our skeleton has the same bones as the skeleton of most other mammals but the relative sizes and shapes of each bone are different for each species. Our spine shows our history in the shape and specialisations of each vertebra. Moving down the spine we have seven cervical vertebrae (pronounced 'serve-eye-cal', from the Latin *cervix*, meaning neck), 12 thoracic, five lumbar and five sacral vertebrae. The sacral vertebrae fuse to each other early in life to form one bone, the sacrum, which forms the back of the pelvis.

YOUR BODY

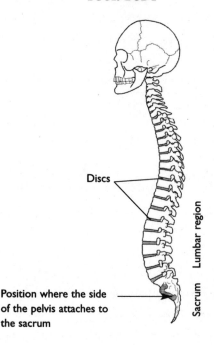

The sacrum gets its name from the same root as 'sacred' allegedly because this sacred bone (in Latin '*os sacrum*') was used in ancient times in sacrifices. At the bottom of the sacrum are three to five small bones which also fuse to each other early in life to form a bone called the coccyx (pronounced 'kok-six' and derived from the ancient Greek word for the cuckoo, '*kokkyx*', because early anatomists thought this bone looked like a cuckoo's beak). The coccyx is the evolutionary remains of our ancestors' tail and we can feel its tip just under the skin between the tops of our buttocks.

YOUR BODY TODAY

The spine today is dominated by our recent move to walking upright. When we were four-legged, the spine was a flexible horizontal bar. It did not evolve to experience compression along its length but when we stood upright this was the result. Now the weight of the upper body presses down the line of the spine. The further down the spine you go, the more weight there is above it. This causes problems for the lumbar region in the lower back (see below). The move to an upright stance was also accompanied by a change in the shape of the pelvis, with the top (the front in other mammals) curving backwards at the same time as a permanent bend developed in the lumbar region of the spine above it.

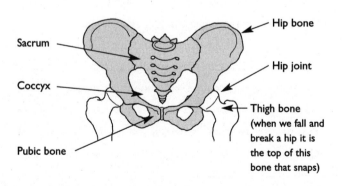

The pelvis of a woman from the front

YOUR BODY

The pelvis is a ring of bone attached to the spine at the rear (the sacrum) and supporting the top of the legs at each side. Our hip bones are the sides of the pelvis and our pubic bone (which can be felt about a hand's width below our navel) is the front. During childbirth, a baby has to pass downwards through the centre of this ring and to allow this the pelvis of women tends to be wider than that of men. Women's hip bones are therefore further apart than men's and hence further from the spine. Women's legs also tend to attach to the pelvis further from the mid-line of the spine, and when a woman lifts one leg when walking she has to lean sideways further than a man to keep the weight of her body directly over the other leg. This is why women and men walk differently.

As the body moved to the upright, it was not just the bones that changed their position; all the muscles attached to those bones had to twist with them. In fact, the complex musculature of the human body is the result of numerous twists and realignments of the skeleton throughout our history: the shoulder bones detaching from the back of the skull; the development of a bendable neck; the legs rotating to move against the side of the body and becoming stronger; the feet being brought under the body with changes to the musculature of the leg joints; ribs disappearing from the lower spine; the primates

acquiring extremely mobile limbs and necks; the head tilting forwards; and most recently the full rotation of the body to stand vertically. This last rotation has put many muscles of the lower back into new positions and this too can cause problems, as we shall see in the next chapter.

Arm

Our arms contain several different types of joint with different ranges of movement. The shoulder is a ball and socket joint, which gives it its wide rotation. It can swing like a propeller in virtually any direction. The elbow, on the other hand, is a simple hinge and only bends in one plane and only on one side of straight. Curiously, some people can swing their elbow past straight and appear to bend it slightly 'the wrong way'. This can look alarming to those of us who can't. The wrist is yet another type of joint. It is a collection of eight small bones and is very mobile, rotating on the end of the forearm. The fingers are each composed of four long bones. The three bones at the end are connected to each other only by two simple hinges but the nearest of the three can rotate on the base of the finger, allowing the finger alone to draw a circle in the air. This base is the far end of the fourth bone, which runs across the back of the hand to connect the finger to the wrist. In the thumb this fourth bone is missing. The thumb has only three long bones but these are like the three end bones of each finger. They are connected to each other by two simple hinges but in this case the base of the second bone is also the base of the thumb. The third bone of the thumb runs inside the body of the hand to connect the thumb to the wrist and, as with the fingers, it is this third bone that rotates on its base. This allows the thumb to swing across the palm and act in the opposite direction to the fingers.

Our wide range of arm movements require a range of muscles to move the bones. As we noted in Chapter 6, muscles often occur in pairs, one to move the bone one way and the other to move it back. These are called antagonistic pairs because they work against each other.

An example of an antagonistic pair of muscles in the arm is the biceps and the triceps. The biceps lies along the front of our upper arm and the triceps lies along the back. The bone between them is the humerus, or 'funny bone' (a play on the Latin word '*humerus*' which just means 'upper arm'). '*Biceps*' in Latin means 'two heads' and '*triceps*' means 'three heads', a reference to the number of anchor points each muscle has. Contracting the biceps lifts the hand; contracting the triceps lowers the hand. If you let your upper arm hang at your side but hold your hand out in front of you, you can feel with your other hand that the biceps on the front of your arm is more tense than the triceps behind your arm. The biceps is contracting to oppose gravity, which is pulling your hand downwards. Similarly, if you rest your hand on the table in front of you or on your leg and push down, you can feel with your other hand that now the biceps is soft and relaxed while the triceps at the back of the arm is tense.

Other common muscles acting on the arm are the pectorals, or 'pecs', of the upper chest which pull an extended arm horizontally across the body (as with a forehand in tennis) and the deltoids of the shoulder which help to extend the arm backwards (like a backhand in tennis). The deltoids are triangular and named after the Greek letter delta ('**D**' – although to look like these muscles it would have to be upside down '**ᗡ**').

Hand

Our hands have already been discussed but it's worth noting they are so adept at gripping branches that, even now, about 4 million years after we left the forest, we surround ourselves with surrogate branches, made to fit the hands we already have. Door handles, bicycle handlebars, car steering wheels, the handles on supermarket trolleys, banisters on staircases, ladder rungs, shovel handles, the oars of rowing boats and a million other examples are all branches in disguise. Anything that needs to be held we make to fit our hand, and our hand was made to fit branches. We even carry suitcases in a simple finger hook. When our ancestors used this grip, the branch was fixed and their body hung from it. With suitcases the shoulder is fixed and the suitcase hangs from that, but the grip is the same and we design the handle to fit our hand like a small cylindrical branch.

Before leaving the hand, there is one whimsical observation we can make. Some scientists claim that, when fingers are measured from the joint at the base to the bone at the tip, men have a third finger (ring finger) that is longer than their first finger (index finger), whereas women have a third finger that is shorter than their first finger. The length of the third finger allegedly correlates with the amount of the hormone testosterone present when the baby is in the womb, and the length of the first finger correlates with the amount of oestrogen. There is more testosterone when the baby is a boy and more oestrogen when it is a girl.

This sounds very improbable, although it certainly works with me. Try asking your friends which of their fingers is longer and see if it's true. (Remember to measure the length of the finger's skeleton; don't just look to see which finger ends further from the

hand. To start the measurement you should feel on the back of the hand, at the level of the knuckle, for the small notch on either side of the base of each finger. This is the gap between the bones.)

Leg

Our legs are dominated by our two-footed stance. Starting at the top, the large muscle of our buttock, the *gluteus maximus* (Latin for 'large buttock') is mainly used for straightening the angle between the thigh and the body as we rise from a sitting position or climb stairs. Weightlifters, who must raise from a squatting position not just their own body weight but the equivalent of several other people's body weight as well, tend to have massive buttocks. This muscle is also used when we lift our leg sideways and it can help pull the thigh backwards when we run but it is not much used in ordinary walking where the angle between the body and the leg barely changes.

The thigh is raised by a number of muscles attaching the thighbone (femur) to the pelvis and spine, whereas most of the muscles in the thigh itself are actually used to bend or straighten the knee. These muscles attach the femur to the bones below the knee through a ligament attached to the kneecap. On the front and sides of the thigh are the quadriceps, or 'quads'. These have four anchor points because they are a collection of four muscles which push the lower leg forwards when we walk, straightening the knee, and also help to straighten the leg when we rise from a chair or climb stairs, or lift unreasonable weights from a squatting position. Weightlifters have bulging thighs too.

The quadriceps are opposed by the hamstrings, which despite their name are three muscles extending down the back of the thigh. These contract to bend the knee and are stretched when the knee is straightened. Athletes and dancers may strain these muscles ('pull a hamstring') when kicking the leg forwards without having properly warmed up. The hamstrings were given their name by butchers who hang the thigh of a pig – a ham – on a hook by passing the hook under the string-like tendons at the knee end of these muscles. You can feel two of your hamstring tendons very easily, especially if you sit down and tense your thigh. They are just behind your knee, one at each side.

The calf muscles of the lower leg are used to point the toes, especially during walking as the foot pushes away from the ground. They are attached to the bone of the heel by the thick Achilles tendon. When they contract and shorten, this lifts the heel and effectively tips the weight of the upper body forwards in front of the toes. We avoid losing our balance by stepping forward, just as we would if someone pushed us unexpectedly from behind. Walking on two legs is just a controlled way of repeatedly not quite falling on your face. When we stand on the soles of our feet, our calf muscles are fairly relaxed but when we raise ourselves onto our toes they shorten and become solid. It is their contraction that is lifting our body.

The calf muscles are opposed by a series of much smaller muscles running alongside the shin-bones. These bend the foot upwards but are much smaller than the calf muscles because they only have to lift part of the foot; the calf muscles have to lift our whole weight.

Feet

Our feet recently had an opposable 'thumb', like those of the chimpanzees and gorilla, but, when our recent ancestors left the trees to walk on two legs, this opposable toe moved back into line with the others.

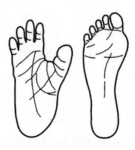

Feet of a chimpanzee and a human

YOUR BODY

A large big toe projecting sideways halfway along the inside of the foot would have made walking with the feet swinging past each other quite difficult. Even today, however, our big toe carries the legacy of its once special relationship to the rest of our foot. It is still possible to bend our big toe upwards while bending the other toes down. No other toe has this independence of movement; the other four always act in unison. It is even possible to pick something off the floor by gripping it between our big toe and our second toe, although this is no longer as easy as it would have been for our ancestors. As a reminder that our other toes were once grasping 'fingers' in their own right, it is also possible to pick something up by gripping it in the curved under-surface of the outer four toes, but it is much harder to do this using only the big toe. This is not surprising. I invite you to try the following trick. Put a pencil on the table then hold your hand palm-down above the pencil. Without using your thumb, and keeping all your fingers together, pick up the pencil by curling your fingers around it. After several attempts you should find this quite easy. You are hooking your fingers in the same way as your ancestors did when hanging from branches. Now try the same thing but keep your fingers straight and pick up the pencil by bending only your thumb. This is much harder (and I still can't do it). Your feet once worked like your hands so you should not be surprised that your big toe still shows its history as an opposable digit but, like your thumb, it is still useless for picking anything up by itself. The main reason for this is that, like your thumb, your big toe has one less joint than your other toes.

Our toes, however, are not just historical artefacts. They have acquired a new adaptation for walking. The end of the four outer toes in humans is double-jointed. The last joint can be bent up as well as down, although – when not walking – we can only make them bend up by taking them in our fingers and bending the tips upwards manually. You cannot bend up the end of these toes by using any toe muscles. This new flexibility has presumably arisen because the tips of our toes are forced upwards at the end of each step. Our finger ends are not double-jointed in this way.

Face

Faces are very complicated three-dimensional objects with many opportunities for subtle differences in their characters (width of nose, prominence of cheekbones, shape of lips). The number of possible combinations of only slightly different features is vast. As a result, it is possible for us to recognise one face out of millions whereas it would be very hard to identify individuals if we were only looking at (for example) their forearms.

The importance of being able to recognise individuals, especially for a visual, social species like us, has programmed the brain in such a way that it wants to find faces, even where there aren't any. I remember being frightened as a child by all the faces in my bedroom, staring at me from the patterns in the curtains and from the floral wallpaper. Worldwide, there are numerous 'Man in the Moon' and 'Woman in the Moon' traditions where people think they have seen a face in the geography of the lunar surface, and numerous books have been written about images of Christ appearing in cloud formations or drifting ice floes. There is also a mountain on Mars which some think looks like the face of a monkey. We just cannot stop seeing faces.

This is such a compulsion that it is possible to see a face with only the barest clues in the crudest form. Cartoons rely heavily on this. An analysis of the faces of cartoon characters in comic strips or animated films will show huge differences between a real human face and the image used, yet we will accept without question, and probably without noticing, that this represents a person. It has even become commonplace in e-mail culture to use punctuation marks as faces, and these don't even have to be the right way up :)

YOUR BODY TODAY

Evolution has not just conditioned us to see faces, it has also given us an emotional response to certain kinds of face. A baby's face is very unlike an adult's. In a baby the eyes are much larger relative to the head, the forehead is rounded and smooth, and the nose and jaw are small. The image of a baby often elicits warm, protective sensations, especially in parents, and again cartoonists (and advertisers) have capitalised on this. Sympathetic cartoon characters are given round faces with outrageously large eyes, quite impossible in nature, while unsympathetic characters are made to look exactly the opposite with long sharp noses, low foreheads and small beady eyes. Our genetic programming is being openly manipulated – beware!

Senses

Evolution has only given us senses to detect those aspects of the environment that were important to our ancestors. Other animals standing beside us may sense a very different world. Hawks can see movement at greater distances than we can; dogs and bats can hear higher frequencies; elephants and whales communicate using lower frequencies; many animals sniff their way through a world heavy with scent; honey-bees and hummingbirds see ultra-violet as a colour; and pit-vipers see infra-red and can detect the body heat of a mouse at 30cm in total darkness (at least, total darkness to us). Some animals even sense the polarisation of light and know the direction of the sun through dense cloud.

Scientists can measure the frequencies of sounds and the wavelengths of colours so they know exactly what our sense organs are receiving, but they cannot do this with every sensation. Pain or taste cannot be measured in the same way, as these are wholly subjective. To some people celery is a refreshing crisp salad, to others it tastes disgustingly acrid and joins a long list of other concoctions that were never meant to be put in the human mouth (blue cheese, taramasalata, hot chillies). Smell is similarly subjective, although the distinction between taste and smell is rather artificial. Both are the detection of molecules which produce a recognisable response. Airborne molecules are detected in the nose, while the mouth senses molecules in a liquid or solid form. Usually both operate together as the smell of what we are putting in our mouths passes up to our nasal cavities (blue cheese). In fact, it has been estimated that 80% of what we perceive as taste is actually being smelled. Holding your nose while swallowing an unpleasant medicine really does seem to work.

We accept that pain and taste are subjective, but are our other senses as objective as we think? We see with our eyes and hear with our ears but it is our brains that take these signals and turn them into perceptions, and we have few ways of measuring what different people's brains perceive. Scientists can measure the wavelength of light reflected from an object and say the object is 'green'. What they cannot do is guarantee that everyone who sees it perceives it as the same green. When one person argues a blouse is turquoise and another says it is green, their brains may genuinely be perceiving different colours, even though their eyes are seeing the same thing. One may see particular shades of green as quite distinct from turquoise while the other sees them as similar and treats those greens and turquoise as the same. Scientific apparatus can tell us what the eye is seeing but not what the brain is seeing, but that problem is beyond the scope of this book.

Eyes

The migration of our eyes to the front of our face during our time as tree-dwellers has given us 140° of two-eyed (binocular) vision to the front – with its accompanying depth perception – and 30° of monocular vision on each side (200° in total). We therefore see blurred objects, or more usually movement, not just directly to the sides but also slightly behind us even when we are looking straight ahead. This is made possible in part by the field of view of the eyeball but also because the sides of our bony eye sockets are cut away and don't act as blinkers. Above the eye, the bone of the skull forms the brow ridge, which differs in everyone but is usually more prominent in men than in women. Below the eye is the cheekbone. These two project forward to about the same level and protect the eye by extending further than the front of the eyeball. When we press our eye to a window it is not the eye that makes contact but our cheek and forehead. The side of the eye socket, however, is further back allowing the eye to see sideways. This is helped by the fact that our eyelids leave a horizontal slit, not vertical, and do not get in the way of our field of view.

Ears

Just as having two eyes allows depth perception, so having two ears allows us to tell from which direction a sound is coming. A noise off to our left will hit the left ear slightly before it reaches the right. To help this direction finding, our ears are as far apart as possible on the sides of our heads, not next to each other in the middle of our foreheads.

The inner ear (used for hearing), middle ear (used for balance) and outer ear (used for sound detection) are all inside the head, but most mammals have an external structure shaped like a cup for collecting and

focusing the sound into the ear canal. This focusing device is made from flexible cartilage (gristle) apart from the human earlobe, which, when present, is fatty tissue. In some mammals, this focusing device can be pointed in the direction of the sound by specialised muscles. Our device – which we usually just call an ear – can't be moved like this, although some people can use the remnants of these specialised muscles to 'waggle their ears'. The only function of this in humans is as a party trick.

In owls, which hunt at night and rely to a large extent on sound, one ear is higher on the head than the other. This increases owls' ability to detect the origin of the noise by allowing them to tell whether it is above or below their line of sight, as well as whether it is to the left or to the right. When people are listening intently they often tilt their head to one side, which produces the same effect, although whether this is subconsciously for the same reason is not clear.

Hair

Although mammals' hair is usually an insulating layer trapping heat, it also acts to block the sun's rays. Selective breeding of European pigs has eliminated their coat of long bristles and they now suffer badly from sunburn unless they can roll in mud or cover themselves in some other substance that acts as a sunblock. Even in apes most of the body is covered by hair, but our bodies have only odd tufts on top of the head, under the armpits and around the genitals. Presumably the hair on our heads is an inbuilt hat to protect the brain from the sun's heat. If it was to protect us against sunburn we would also expect to see it on top of our ears, nose and shoulders where we are burned most. Also, natural selection would only have given our ancestors a defence against sunburn if sunburn reduced their chances of survival, which is by no means certain.

The hair in our armpits and groin is associated with concentrations of sweat glands. This hair is crooked and long and, unlike the hair on our heads, resists lying in a dense mat. This seems to be the body's way of maximising the surface area available for dispersing the natural scent produced by the glands.

The hair on top of our head, and on the jaws of men in some racial groups, continues to grow throughout our life, getting ever longer unless we cut it (although in some parts of Africa the hair on the head is naturally brittle and never reaches a great length before it breaks). Hair that keeps growing like this is very unusual for a mammal, and seems to have evolved only recently. However, now most of us have long hair on our heads, we can shape it as a signalling device and in most cultures people spend huge amounts of time and money doing just that. Many men also undergo a daily ritual of scraping the hair from their faces with a sharp metal blade. This removal of adult facial hair growth is presumably also a signal, although the message is not clear. Mercifully, the tufts of hair on the rest of our body stay naturally short, as do the sparse hairs on some men's arms, legs, chest and occasionally back.

Eyebrows are our other main tufts of hair. The reason we have eyebrows is not wholly obvious but they are presumably there as a result of natural selection and would be expected to have some value to survival or the enhancement of childbearing. At school I was taught eyebrows have evolved to keep the rain out of our eyes. This is not a particularly convincing argument (and not least because I have many times suffered from having rain in my eyes). If anything, they are better at preventing perspiration rolling into the eyes, but is this any better as an explanation? Is it credible that in a social species the only individuals to survive and breed were those with bushy eyebrows that kept the perspiration out of their eyes?

Another suggestion is that eyebrows arose as signal flags to accentuate the eyes in social interaction. It is common in primates to 'flash' the brow ridge upwards and we certainly use this facial gesture ourselves, especially when passing the same person for the umpteenth time that day, when the 'hello' and smile of the first encounter later wanes to a nod of the head and eventually ends up as only a flick of the eyebrows. This eyebrow flash is virtually universal in all human cultures. Unfortunately, if this was the eyebrows' function, they do not seem to be very good at it. Sixty thousand years ago, everyone had dark skin and dark hair (and presumably had eyebrows). If eyebrows were a way of accenting the movement of the brow ridges, there would have been advantage in selecting a colour that contrasted with the skin. This has not happened in today's African peoples and presumably did not happen in everyone's African ancestors.

Some people have suggested eyebrows are important because they diffuse the glare of the sun. This is possible and if true would suggest eyebrows evolved after our ancestors left the shade of the forest. Our closest relatives the chimpanzees have prominent brow ridges but not obvious dense bands of hair as we have.

The truth is we do not know why we have eyebrows. They can trap perspiration; they can accentuate the eyebrow flash if they are dark on a pale skin; and they may diffuse glare from the sun, but that does not mean they survived (or appeared) for any of those reasons. What we can say is that nowadays, especially in industrialised nations, they are treated primarily as cosmetic adornments. In some cultures, women will regularly pluck the eyebrow hairs from their faces one at a time with metal pliers until only a narrow line remains, then restore the original effect with a black pencil which is reapplied daily. The origins of this practice are obscure.

Male-pattern baldness

Despite many claims to the contrary, especially on the internet, the causes of male baldness are not understood. It is found especially in Caucasian men and there appears to be some genetic influence but young men whose middle-aged father shows no tendency to baldness should not be too confident they will follow in his footsteps. My father died aged 81 with a full head of hair but I started to lose mine at 25. On the other hand, my mother's father was bald.

Whatever the cause of male baldness, it usually only starts to make an appearance after the early childbearing age of our recent ancestors. Historically, it would therefore not have been a factor when women were considering potential husbands – but would that have mattered? Does baldness make men unattractive to women? Well… despite what women tell men (as distinct from what women tell each other), the answer is possibly yes. 'Must be bald' is not usually high on any woman's list of desirable characteristics for a future partner. Having said that, such lists usually have imaginary men in mind but our female ancestors would have found their partner among the real men they knew. Women may not dream of bald men, but some do marry them.

Heat control

The ability to control our body heat is one of our primary mammalian features. We have a normal core temperature of about 37°C (98.6°F), although it can be up to 0.8°C higher or lower than this depending on the individual, and when we reach old age our normal temperature is lower than it was when we were younger.

If our internal temperature rises to 40°C or more, we may suffer convulsions, coma, brain damage and death. If it falls below 35°C, we can suffer mental confusion and slurred speech. Below 30°C, there will be lowered blood pressure, pulse rate and respiration, and eventually, at about 27°C, death due to hypothermia.

Through evolution, we have lost most of our insulating layer of hair but for keeping warm in cool climates we have gained clothes, fire, insulated houses and central-heating boilers. In hot climates we can

YOUR BODY TODAY

keep cool with light clothing, cold showers, cold drinks, fans and air conditioning. Without these technologies, our body has various inherited mechanisms for cooling or warming itself.

We lose heat by sweating, by panting and by expanding the blood vessels just under the skin to radiate more heat. These swollen capillaries give a reddish tinge to the skin and make us look 'flushed'. Our naked skin has more sweat glands than any other primate, which suggests our body's hair loss was also a way of losing heat.

We generate heat by exercising, by shivering and by contracting the blood vessels just under the skin to slow the heat loss. These narrow capillaries make the skin less red, especially over the bones of the face, and in Caucasians this leads to the expression 'turning blue with cold'. Jumping up and down, stamping our feet or waving our arms are all just ways of making our muscles contract, which generates heat as a by-product. In shivering, the body does this for us. Shivering is a fast involuntary vibration of the muscles to generate heat.

Our ancestors could also conserve heat by making their fur stand up to trap a thicker layer of insulating air. We still have this response but not the fur. Our skin has only sparse or minute hairs but these still have the tiny muscles that pull each one upright.

When these muscles contract in cold weather, they produce the familiar 'goose-pimples' or 'goose-bumps' (which are no help at all).

Body shapes

Once men and women reach sexual maturity at puberty, their bodies change. At this time the bodies of boys and girls change in different ways. Boys acquire much more muscle and become noticeably stronger, whereas girls acquire proportionally more fat. This is an energy reserve being stored against the day when they are pregnant and must nourish a child even if environmental conditions are harsh and they are unable to feed themselves. This extra fat in women and extra muscle in men accounts for much of the difference in the shape of men's and women's bodies.

In addition to these gender differences, there can also be differences in the overall shape of the body in races originating in different climates. Peoples who recently evolved in very cold climates tend to be stocky. In all people, heat is lost from the body surface and being stocky reduces the amount of surface relative to the body mass. This is geometry. The shape with the smallest surface for any given volume is a sphere. Those people who get into the record books by writing the Lord's Prayer or some other body of text on a grain of rice really do deserve our respect;

a grain of rice is quite close to a sphere and has a very small surface area for its size. If they boiled the grain of rice and squashed it flat they would find the job much easier because the surface would now be much larger. Conversely, the thin flat page in your hand would have very few words on its surface if it was screwed into a tight ball, but the volume would not have changed. In a warm body, such as an animal, heat is lost mainly through the exposed surfaces. The closer to a sphere it is possible to get, the less heat will be lost because the surface will be relatively small for the volume. This is why cats and other mammals curl into a ball when resting during cold weather, and why animals which hibernate also curl into as tight a ball as possible.

On the other hand, peoples who evolved in equatorial regions tend to be slim and may be very tall, for example the Dinka of southern Sudan. This body shape reduces heat stress, especially during exercise, by providing a large surface area relative to its volume and hence increases the amount of heat radiated away.

Disposable organs

Throughout our lives, some of us will have one or more of our organs surgically removed, usually (or hopefully only) as a result of illness. In some cases, this will drastically alter the way we live. If our kidneys are removed, we must substitute them with a machine and undergo regular dialysis to clean our blood or we shall die. If our gall bladder is removed, we must alter our diet to avoid fats (the gall bladder passes bile into the intestine which helps break down fatty food). However, there are some organs that can be removed with virtually no noticeable effect on our long-term health or our everyday lives. The most obvious examples are our tonsils and our appendix.

Tonsils

Our tonsils are at the back of our throat, one on each side, and are part of the body's defences. They contain cells belonging to our lymphatic system, which is a network of tiny channels spreading throughout our bodies, virtually wherever there are blood vessels. These channels drain into 'lymph glands' or 'lymph nodes' in the neck, groin and armpits. The lymph system exists because blood is pumped around the body under pressure – our 'blood pressure'. Because the blood vessels are pressurised, it is perfectly normal for some of the blood fluids (not the cells, just part of the plasma) to seep out of the blood vessels into the surrounding tissues. If this 'lymph' was left to build up, it could cause a problem. It's the job of the lymphatic system to drain this fluid back into the blood near the heart, but while it does this it also monitors the fluid for disease-causing agents. If these are found, it mobilises special cells from the body's defences to attack and destroy the invaders. This is why an infection can sometimes cause the lymph nodes to swell. Doctors feel the sides of your neck during fevers in part to see if the lymph nodes are swollen. A swelling indicates that the body's defences are actively fighting a problem.

Tonsils contain tissue belonging to this lymphatic system and monitor the condition of our food as we swallow it, checking for harmful contamination. With this role, tonsils may become infected and swollen, producing tonsillitis (the ending '-*itis*' means 'inflammation'). In the past, troublesome tonsils were usually surgically removed, but this is not always the case today. Even if they are removed, there are other parts of the digestive canal that also monitor the condition of our food and these continue to alert the immune system.

Appendix

Our appendix can be removed without side effects because it is one of our body's remnant organs, the evolutionary remains of a structure used by our distant ancestors but mainly redundant in us, like our tails.

YOUR BODY

The appendix is a small blind-ended side branch of the intestine. In some other mammals, for example rabbits, this is a large offshoot of the intestine called the caecum (pronounced 'seekum', Latin for blind) in which tough plant material is held while its relatively indigestible components are broken down by helpful bacteria. Our far-distant ancestors apparently used this process, but in humans the caecum is no longer active in digestion and has shrunk to only 9cm, about the length of a finger. By comparison, our whole digestive system, from end to end, is 9m long.

Although no longer active in digestion, the appendix is not wholly without a function. It too is part of the intestine responsible for monitoring the condition of what we swallow and for detecting harmful foreign bodies. Like the tonsils, it is not unusual for this small cul-de-sac to become infected and swollen ('appendicitis') and it is often removed by surgery. Again, this produces no obvious harm to the body as other parts of the intestine also monitor our food.

The appendix is a part of the body which has become redundant during the course of evolution, but some of us carry two structures which became redundant during our development in the womb. These are men's nipples.

YOUR BODY TODAY
Why do men have nipples?

Correctly, it is not nipples that are obvious on men's chests, but the patches of pigmented skin around nipples, known as areolae (Latin for 'little areas'). The presence of such obviously useless items in men results from the way our bodies acquire their sexual identity when we are embryos.

Perhaps surprisingly, all embryos start to develop in the womb as females, regardless of whether they are genetically female (with two X-chromosomes) or male (XY). We all began life as women. During this early stage of development, the cells for the production of breasts and nipples form in the chest region of every embryo. However, if the foetus is XY, about six weeks after fertilisation, one of the genes on the Y chromosome springs into life and triggers the formation of testes. These then produce testosterone – a steroid hormone produced by the testes, hence the name. Under the influence of the testosterone, the XY foetus changes to become male. Without testosterone it would continue to develop as a female. (Testosterone is an example of an 'androgen' hormone, from the Greek '*andro*', meaning 'of man', and *gen*erator.) However, even with testosterone it is too late to remove the cells that will later form the man's areolae. This is why men have nipples.

In some rare cases, the body of the XY foetus does

not respond to the production of androgen. Testes form inside the body and the hormone is produced but the body simply ignores it. This is known as Androgen Insensitivity Syndrome (AIS). In this case, as all foetuses start life as female, the foetus simply continues to develop as a girl, despite being genetically XY. The AIS may be complete or partial, but if complete the resulting woman may be no more masculine than any XX woman. Indeed, even XX women will produce some androgens in their bodies which may have some small effect, whereas the body of an XY woman may have no reaction at all to androgens. There is one problem however for an XY woman: she will be unable to have children. In an XX foetus, the sex chromosomes cause the development of ovaries. These produce the female oestrogen hormones (from the Greek '*oistros*' meaning 'madness' – a reference to the activity of many animals during their breeding seasons – and *gen*erator). Oestrogen triggers the development of the uterus. At puberty, the uterus takes the adult form and renews its lining every month, shedding the old tissue in menstruation. An XY woman has simple internal testes, not ovaries; consequently, she does not produce eggs, and her fallopian tubes and uterus do not form and she can never menstruate. Today, it is sometimes a medical examination to determine the reason for the lack of menstruation that leads to the

discovery she is genetically XY. Other than that tragic consequence, however, she is fully female. Without men's bodies reacting to testosterone in the womb, we would *all* be female. Female is the default human condition, as the nipples on men's chests testify.

Behaviour

We have already considered inherited behaviours in newborn babies – crying, suckling and grasping – and our fear of the dark, but there are other inherited behaviours that are universal for the human species. Smiling, laughing and frowning have been found wherever there are people and they seem to mean the same things to everyone. Other primates, however, do not share these behaviours. Chimpanzees and gorillas both view a display of teeth as aggressive. It is not a good idea to grin at a gorilla.

Breeding is also an inherited behaviour for all species. Most animal species have a distinct breeding season related to the time of year and regulated by their hormones. We are one of the few species that does not have a breeding season. Humans can produce children at any time of the year.

To sleep: perchance to dream

Sleep is a behaviour that evolved very early in our evolution. In fact, it is rare for any animal species to be active 24 hours a day. Even many marine fish appear to sleep; they become still, rest on the seabed and seem to be unaware of their surroundings. At night, scuba divers can gently pick these fish up in their hands, but if handled too roughly they appear to wake and race away as if startled.

In humans, sleep is thought by some scientists to be important for the immune system and it may be a recharging period. The function of dreaming is less well understood and it is not clear how widespread it is in the animal kingdom, although anyone who has ever watched a dog sleep can have little doubt that dogs share this ability with us.

Sleeping and dreaming is potentially a dangerous combination. If we started reacting to the virtual-reality world of a dream as if it were true reality, we could harm ourselves. Our ancestors could have wandered into the path of nocturnal predators or fallen out of trees while asleep if the body was not immobilised in some way. Natural selection achieved this immobility by evolving a form of 'sleep paralysis'. When we sleep, our muscles are effectively disconnected from our brains so that, when we dream that we are walking or running, our bodies are actually relatively still. It is not difficult to see how this could have evolved. Any animal with genes that failed to immobilise its body would probably not last long enough to pass those genes to any offspring.

Sometimes in humans, this sleep paralysis breaks down and people sleepwalk. This is a complex phenomenon as the sleepwalker appears to be asleep yet interacting with the real world, not with an imaginary dream-world. They walk around furniture and pass through doors, and have their eyes open.

Sleep paralysis sometimes lingers into a state where we are half-asleep and half-awake, when it can be quite terrifying, as any sudden awareness of paralysis would be. In medieval Europe, people who experienced this believed an evil spirit was sitting on their chest while they slept, pinning them to the bed. They called these spirits *Mæres*, which is the origin of the word 'nightmare'. A space-age interpretation of this sensation may lie behind some people's belief that they have been visited in their homes by extra-terrestrial beings who immobilised them in their beds. I have certainly experienced a related sensation while half-asleep in which I had the distinct impression that my whole body was paralysed and being pulled out of bed feet first while my legs sank downwards into the mattress to an angle of nearly 45°. Fortunately, the sensation evaporated when I managed to wiggle my toes (obviously a gesture terrifying to aliens).

Chapter Thirteen

Your Body's Problems

The tortuous evolutionary route to our current bodies has left us with a number of physical problems, the most common of which is back pain. This can manifest itself in several ways but is often caused by careless posture when lifting heavy objects.

The lumbar region of our spines, at the level of our waist, can still flex sideways like a reptile's and forwards and backwards like a four-legged mammal's. Despite the fact that each vertebra has interlocking projections which resist excessive movement, this is still a very flexible structure. It also has a permanent curve, even when we stand upright, and has only recently – geologically speaking – been subject to compression along its length. This could hardly be a worse combination of features for a structure that is to be put under a heavy load.

For this reason, it is imperative when we lift heavy objects to keep the spine as vertical as possible. If we lean forwards or sideways, keeping the spine bent, or twist at the waist, one of two serious problems can occur. First, we may suffer a 'slipped disc'. Discs are circular pads that sit like washers between each of the vertebrae. They are slightly compressible (we might say rubbery) and act as shock absorbers and flexible joints. They have a tough outside coat but a core like stiff jelly. If the spine is bent and then put under a load, two vertebrae may squeeze unevenly on the disc separating them, pinching one side and forcing the opposite side to bulge outwards (like squeezing a cherry hard between your finger and thumb until the stone pops out). A bulge like this can press on the nerves around the spinal cord causing pain in the lower back and legs. Treatment for a slipped disc usually involves rest, anti-inflammatory drugs and perhaps physiotherapy, but sometimes surgery may be needed to release the pressure on the nerves.

The second potential problem with the back is muscle injuries. If the spine is kept as straight as possible when lifting weights it can take some of the

YOUR BODY'S PROBLEMS

load, like a pillar. If it is bent, much of the loading passes to the muscles which then strain to keep the body upright. There are many muscles in the back and they run in different directions. Under an unbalanced load, it is not difficult to cause damage.

Back muscles can also be damaged in less obvious ways. A consistently bad sitting or sleeping posture, especially if a bed is too soft, can strain the muscles of the back and cause problems that recur for years.

Once back muscles have been damaged, the slightest attempt to move causes intense spasms of pain. This makes walking, or even standing, impossible until the swelling subsides, which may take days. Even then, the muscles may never fully recover and the back may be left feeling permanently weakened. Back pain is one of the main causes of absenteeism from the workplace. We should always adopt safe postures for sitting, standing and lifting.

Our lower back causes these problems because for most of its history it was not shaped for an upright posture. This history also causes a problem at the other end of the spine – whiplash.

Normally, a mammal holds its head on the front of a horizontal backbone. If the animal is struck from behind everything is in line and the sudden acceleration of its body is absorbed along the length of the skeleton. Several features have changed in humans: our spine is vertical; we have evolved a large brain which is heavy; we have a very flexible primate's neck; and we have invented large heavy solid vehicles which we propel at unnatural speeds until they hit each other or people.

When a vertical person (sitting or standing) is struck unexpectedly from behind, their spine can no longer absorb any acceleration forces and is propelled forwards. Meanwhile, their heavy head tends to stay where it is. Because the human neck is so flexible, it cannot counteract the difference in movement between the spine and the skull. Instead, it acts like a hinge, allowing the head to flip suddenly backwards. This can crush the neck vertebrae and seriously damage the muscles and nerves.

Heart

Our recent move to a vertical posture means our heart now has to pump blood uphill against gravity, not horizontally, to reach the brain. Walking on two legs has also made those legs large and muscular. At any one time, the legs contain a significant proportion of our blood, all of which must be returned to the heart against gravity.

In times of stress, it is therefore advisable to change our posture. When we feel faint, we can put our head between our knees to let gravity take blood to the brain, and if someone is unconscious we can put them in a coma position, horizontally on their side, again to minimise the heart's need to pump uphill.

Teeth

Throughout our time as primates, our jaws have become smaller and our large canine teeth have shrunk. We now have a much flatter face than our ancestors. Unfortunately, natural selection appears to find it easier to change the shape of something than to alter how many somethings there are. Some of our teeth may be smaller but we still have 32, the same number as chimpanzees and gorillas. As these teeth are now housed in significantly smaller jaws, it is not unusual for them to be overcrowded and pushed into unnatural positions. Many people today spend at least some of their childhood wearing metal bracing designed to force their teeth back into line.

The last teeth we grow are the four molars at the very back of our mouths, the so-called 'wisdom teeth'. By the time they try to appear, there is often so little space left that they become jammed under the teeth next to them. These are 'impacted wisdoms'.

Hernias

A hernia, or rupture, is when one of our organs, or part of the intestine, pushes through an abnormal hole in our body, usually through or between the muscles of the abdominal cavity. This type of injury can happen if the pressure in the abdomen is suddenly raised by a violent cough or by lifting a heavy weight.

One of the commonest types of hernia occurs because we are mammals. When a man's testes move out of the body cavity and into the scrotum during development, a weak spot is left in the abdominal wall on either side of his groin. If he later exerts himself in an unfortunate way, this weak spot can rupture and part of the small intestine can be forced through the hole to appear as a painful lump under the skin.

The usual remedy for this is a simple surgical operation to replace the intestine and repair the hole. In extreme cases, however, the hole can tighten around the protruding loop of intestine, trapping it. This is a 'strangulated hernia' and is more serious as it may cut off the function and circulation of the trapped bowel.

Various other types of hernia exist in both men and women. To avoid them, it is best to avoid lifting excessively heavy weights and any activities that may cause sudden increases of pressure in the abdomen, as when the abdominal muscles are suddenly clenched.

YOUR BODY'S PROBLEMS
Wearing out

In all but the poorest communities, most of us live much longer than our ancestors thanks to better living conditions, less arduous lifestyles and more effective medicines. Unfortunately, like some manufactured products, many components in our bodies seem to have 'best before' dates. As our ages exceed these dates some of our parts begin to wear out.

This is especially a problem with the joints of our legs. Again, this results from our evolutionarily recent move to an upright stance. Each of our hind legs spent most of its history adapting to share the weight of the body with three others, or two others when one leg was stepping forward. We now carry the full weight of our bodies on two legs and as we walk we are balanced on one leg which is changing position and rotating its joints under this load. We have also grown larger and heavier over the last few million years and for many of us our new urban lifestyles, with their fatty convenience foods and lack of exercise, have exacerbated the weight problem. These factors have all taken a toll on our knees and hips.

The ends of the bones in our joints are covered by a smooth layer of cartilage bathed in a lubricating fluid which reduces friction. As we age, the cartilage layer can wear away leaving the ends of the bones in direct, grinding, contact. This is known as osteoarthritis (literally 'bone joint inflammation'). Osteoarthritis can also occur between the vertebrae of the neck or in the lumbar region of the spine, and in the fingers and thumbs.

In extreme cases, and if medical facilities are available to us, surgery may relieve the symptoms. In the spine, it may be possible to fuse the offending vertebrae together. This reduces flexibility but stops the grinding of the bones. To restore movement to the legs, the ends of the bones in the knee or hip can be surgically removed and replaced with a metal and plastic alternative. These are never as strong as the original but they greatly improve the condition for the patient.

Our muscles also weaken as we age but these never seem to wear out in the same way as bones, especially if they are used regularly. No one has ever lost the ability to talk because their tongue has worn out.

It is not just our joints that deteriorate with age. Our teeth can wear away and our sense organs can lose their acuity. Again, we are the only species that consciously addresses these problems with technology – dentures for chewing; spectacles, contact lenses or laser surgery for seeing; hearing aids (or the less socially acceptable volume controls) for hearing. Technical and surgical interventions like this are impressive examples of our species' unique capacity to imagine the most outrageously bold solutions to nature's inconveniences, then find a way to bring those solutions into existence. Organ transplants, blood transfusions, brain surgery and numerous other examples we now take for granted are all rooted in our ability to take leaps of imagination almost certainly not possible for any other species in the last 3,500 million years. Who, a thousand years ago, would have dared to dream that we could save a life by taking the heart out of an accident victim and stitching it into someone else, or by draining all the blood from someone's body, laundering it, then giving it back to them while they sit and read a book? We truly are an amazing species.

Ageing

The reality of ageing is a more general problem than just the wear and tear on our body parts. As we age, our immune system becomes weaker, digestion becomes less efficient, lungs become less elastic and it gets harder to breathe, the heart is less efficient and arteries become thicker and less elastic, temperature regulation may become less efficient and, with poor circulation, can lead to our feeling cold even in warm weather. Loss of nerve cells in the brain affects memory, the kidneys become less efficient and the bladder becomes less elastic requiring more trips to the bathroom, especially during the night. Externally, our skin becomes thinner and wrinkled, hair may turn grey or fall out, our muscles lose their tone, our faces sag and it becomes harder to lose unwanted fat deposits, especially around the waist. Our ears become longer and thicker and in men they grow long embarrassing hairs, especially in the entrance to the ear canal. Bones become increasingly porous and may collapse slightly in the spine, adding to a natural thinning of the discs between the vertebrae, causing the whole body to shrink and stoop.

Scientists have suggested various reasons for this catalogue of woe. Some say that cells may have a limited active life – that they actually (not metaphorically) have a 'best before' date. Once their life is completed their genes deliberately destroy them. This is an extension of what happens naturally in many tissues even early in life, with cells dying and being replaced. It occurs in the skin where cells die as they move towards the surface to form a protective outer layer which is constantly worn away (most house-dust is dead skin cells). The suggestion is that later in life this cell suicide happens to all our cells.

So-called 'free radicals' (electrically charged molecules) have also been proposed as a cause of ageing. Free radicals are created in the cells during normal activity as chemical processes release energy.

They are the cell's chemical pollution and if they are not neutralised they can be very damaging. Early in life, the body produces natural antioxidants to counter them but later becomes less efficient at this and the free radicals become increasingly destructive, damaging the genes that oversee correct cell function. Some people now think this damage is the main reason for ageing.

Ageing and genes

Some scientists have suggested that ageing is the result of an accumulation of genes that have a negative effect on the body – perhaps by reducing the effectiveness of antioxidant activity – but which only come into play later in life.

Not all genes are active from the moment of the fertilisation of the egg in the womb. Some lie dormant until later in life when they are said to 'switch on'. Puberty, when the body changes from that of a child to that of an adult, is a good example of a time when genes which have been lying dormant switch on and tell the body to produce the hormones which bring about maturity.

All our genes first came into existence when they evolved from older genes. Genes change with time (they 'mutate') and these changes may not always be beneficial to their owner. However, even harmful genes may survive in the population; it all depends when they act on the body.

Any newly mutated gene that switches on before its owner reaches breeding age and then inhibits the number of children they can have may die out quickly because it is not easily passed to the next generation. Conversely, a gene that switches on before breeding age but does not inhibit the number of children, or perhaps enhances the number of children, may well pass to the next generation. However, any gene that switches on later in life will *always* pass to the next generation, because the parent has already passed it to the children before it has switched on and had an effect – whether good or bad. Therefore, even destructive genes will always survive if they just happen to be a gene that switches on later in life. Some people have suggested ageing may be the result of accumulations of this type of gene.

Ageing and natural selection

If some genes can survive despite being highly destructive to their owner, it is clear natural selection is not doing their owner any favours. This is true, it isn't. The reason why is best explained by another analogy.

Evolution is like a relay race where the genes are the baton. In a relay race, the commentator concentrates on the athlete with the baton, but only until the baton is passed to the next runner. As Runner 1 sprints round the track, the commentator focuses completely on their efforts, but when Runner 1 passes the baton to Runner 2 the commentator does not continue to describe Runner 1: how they slow down; how they are panting; how they are bending forward and resting their hands on their legs. Instead, once the baton has been passed to Runner 2, the commentator switches their whole attention to Runner 2, the runner with the baton. The commentator does not care if Runner 1 has collapsed, the focus has passed. Evolution is the same. Once the genes have passed to the next generation and that generation has become independent of the parents, evolution does not care whether the parents fall over or not. The parents become expendable, of no consequence. They have already achieved their success. They have passed the baton. At this stage, there is no need to keep the parents young and healthy. Natural selection has deserted them in favour of their children.

Of course, if we don't have children, we still age, and we still age at the same rate as parents (although parents may feel they age faster). To continue the analogy, if we don't have children, we are the runner who fell and didn't finish the race, taking the baton down with us. Natural selection, like commentators, ignores runners who have fallen.

We said in an earlier chapter that childbirth was painful simply because it doesn't have to be painless. We can add that our bodies age because they don't have to stay young. Like a plant that has already shed its seeds, its job is done. It doesn't matter to natural selection if it now withers and dies.

Chapter Fourteen

Your Brain

Homo sapiens is just another animal species but we are one of those rare species, the generalist. We are not locked into one narrow biological niche in the way many species are. We do not rely on one food source; we are not restricted to a narrow temperature range; we do not live only on mountains or only in deserts. We can exploit virtually any environment and any food. When our ancestors evolved in Africa, they were not ideally suited to the environments of Lapland or the Amazon basin, yet we have thrived in these places because our intelligence has allowed us to solve the problems we faced and to mould our lifestyle to the conditions we encountered, making the clothes we needed and taking our food from what was available.

Our adaptability has made us remarkably successful at perpetuating our genes and there has been a massive increase in our population and its dispersal across the planet. We are now so numerous and our technology so powerful that we are having a serious effect on our environment and the other species in it. If a plant had the characteristics of *Homo sapiens*, it would be seen as a pernicious and undesirable weed. We are the Earth's greatest weeds.

Perhaps this is because we haven't evolved as a global species. We have become a global species only recently and don't yet have a global outlook. 'The World' is over the horizon for most of us – and for some more than others. We focus on our surroundings and on meeting the goals of our local culture, not the needs of our international culture. We want global warming to stop but we won't stop using our cars. We want to eliminate world suffering but we celebrate 2,000 years of Christ saying so by spending £800 million erecting a large tent in London. We are by nature a parochial species; mentally, we have not escaped the village and our intelligence tends to be used to gain knowledge, not to address what it tells us.

Homo sapiens 'Wise man'

We certainly think of ourselves as the most intelligent species on the planet. Even if we believe our bodies are only slightly different from that of a chimpanzee, most of us will consider that our superior intelligence is the one evolutionary achievement that sets us apart from all other life-forms.

We all understand what this claim means, and if we are asked to define 'intelligence' we say it denotes 'cleverness' or 'brain-power' or any number of similar terms. The problem is, these are not definitions, they are just other words we use to mean the same thing. None of them actually explains what that thing is. If we are the most intelligent animals on the planet, we should be able to decide what 'intelligence' is, but this is not as easy as it sounds. Whole conferences have been organised to consider intelligence and hundreds of brainy people have gone home at the end none the wiser. It has even been suggested we should stop talking about intelligence as definitions are virtually impossible, but it's already too late for that. We have the word, everyone uses it, and we all seem to use it in the same way. This suggests we all think we know what it means, even if we can't say what that is.

If we had to choose an analogy for intelligence, most of us might choose something like 'strength'. Strength is something we all have to differing degrees and people can be compared and ranked from very strong to very weak. Strength is usually how we think of intelligence – brain strength – but is this analogy valid? Perhaps we can get some clues by looking at the way we use the word.

The nature of intelligence

There is no dispute that intelligence is only a property of our brain. A surgeon cannot remove our intelligence and study it under a microscope. There is also no dispute that, if someone tells us we've got some, it's definitely a compliment. We approve of intelligence. We also seem to view intelligence as a relative quality. We measure it by comparing people with each other, or comparing other species with people. If other people do things with their brain we cannot do, we say they are intelligent. If other species behave as people do, we say *they* are intelligent (but we never forget we have more intelligence than they have).

We assess the intelligence of other species by designing tests to which we know the answer and judging how closely or how quickly an animal that does not know the answer gets to it. This is very egocentric and rather arrogant. It is simply not valid to assess how intelligent a chimpanzee, a dog or a whale is by using this criterion. This approach is rooted in the outdated view of 19th-century naturalists who saw evolution as a ladder climbing progressively closer to perfection, with ourselves sitting on the top as the most advanced and intelligent species of all. They then attempted to rank other species to see how far up the ladder they came.

Unfortunately, this view of evolution was flawed. It is not a ladder. Every species alive today has spent the same length of time evolving as every other, and, as each is clearly a survivor (it is alive today), it is presumably as intelligent as it needs to be. It is wrong to see other species as having travelled the same evolutionary path we have but having stopped before they reached the end. Each has travelled its own path. We cannot compare the intelligence of a dog with the intelligence of a whale. A dog is as bright as a dog needs to be in its world and a whale is as bright as a whale needs to be in its world. Attempting to compare them is like asking, 'Which is better, an oak tree or a herring?' Better for what? They have evolved for different lives in different environments. Comparing them is not comparing like with like and attempts to do it show a lack of understanding about evolution and ultimately about nature.

What we exclude from 'intelligence'

In the second chapter, I said two cornerstones of the human condition were our logic and our intuition, and that science reflected our logic and religion reflected our intuition. In fact, there is far more than intuition and religion to this second aspect of our identity; this is also our moral and ethical sense and it is often in conflict with what we do with our science. The question we ask of science is: 'Can we…?' ('Can we make atomic bombs?'; 'Can we live without eating meat?'; 'Can we clone human beings?') It is the non-science side of human nature that asks, 'Should we…?' If anything, this is a far more important question. It is not our unique genius at manipulating nature that distinguishes us as human, it is our ability to ask, 'Should we?'

We can devote our technological resources to landing a probe on Mars while in parts of the world children die for want of clean water or from easily treatable infections. We allow this because they are other people's children and they are not dying in front of us. We know these children are there and that their plight is undesirable, yet we choose to do little or nothing as a species to help them. Our capabilities may convince us how intelligent we are, but our choices seldom support this self-congratulatory definition.

Nevertheless, we talk of our 'humanity' and our capacity to be 'humane', words drawn from our nature as 'human', but these words do not refer to our intellect. We choose to define ourselves by acts that come from our heart not from our head.

Similarly, we say Einstein – a prominent scientist –

was intelligent. Yet we say Rembrandt – a prominent painter – was talented. We do not consider great artists intelligent (I don't suggest artists lack brainpower, only that we don't require it of them). This 'Arts and Sciences' distinction seems to be deep-rooted. It reflects, again, the different facets of what we perceive to be human. We describe a work of art as 'having heart' or 'having soul' but we don't describe it as 'having brain', and, when we discuss that vague yet quintessentially human idea of a soul, it invokes, again, expressions such as 'heart and soul' not 'mind and soul'. Do we perhaps consider mind to be the mere manipulation of knowledge?

Knowledge and intelligence

When we think of people we consider intelligent, they are seldom individuals who lack knowledge. They may have a vast store of academic knowledge, like a rocket scientist, or they may have knowledge of the current circumstances, like a diplomat trying to avert a war. Some knowledge seems to be needed for intelligence. On the other hand, a vast store of knowledge is not enough for intelligence. Even the largest computer database in the world is not considered capable of intelligent thought, and someone who knows everything there is to know about Norwegian postcodes may still not be considered an intellectual by their friends (in fact, probably the opposite).

Conversely, some people can be hailed as intelligent while having limited knowledge. Child prodigies have not lived long enough to acquire knowledge, yet they are brilliant in their fields. Their fields, however, tend to be of a certain type. Child prodigies tend to excel at subjects like mathematics, music or chess. These subjects are complex because of the layering of essentially simple constructs. Mathematics is just addition and subtraction, and even subtraction is just addition backwards. Music is just a scale of sounds, some of which clash and some of which harmonise. Chess has a small number of pieces, each of which can only move in a particular way. The challenge of these subjects is the infinite number of ways these simple elements can be combined to produce different results. This resembles the way a skilled bricklayer can take many identical bricks and a supply of mortar and combine them to produce many different types of house. The raw materials are limited; the skill lies in the ability to manipulate them. There are few child prodigies in medicine, architecture, engineering and a million other subjects because these disciplines require knowledge – vast banks of raw data. A six-year-old child hasn't acquired knowledge. Prodigies have acquired skills for certain subjects easily because of the way their brains work compared to other people's brains but they have not had the time to acquire data.

So, knowledge may be related to intelligence – we even use the word 'intelligence' to mean 'information' (Military Intelligence, Secret Intelligence Service) – but knowledge alone is not the same thing as intellectual intelligence.

Curiosity and intelligence

We are probably the only species that asks, 'How are stars formed?', 'What is inside an atom?' and 'How can continents drift across the surface of a solid Earth?' This is a function of our species' strong sense of curiosity. Curiosity is inherently a dangerous thing – and would have been especially so in the untamed world of our ancestors – but the rewards of curiosity can be considerable. Curiosity is a thirst for knowledge. When we say someone who asks 'Why?' is posing an intelligent question, we are not recognising their curiosity as a form of intelligence, but we are subconsciously acknowledging a connection between knowledge and intelligence and expressing the view that it is intelligent to seek knowledge.

Language and intelligence

We can only ask each other, 'Why?' because we have developed spoken, and more recently written, language. Philosophers are fond of saying that the human development of language is, above all else, an indication of our intelligence but whether this means anything depends on whether we can solve the riddle of what intelligence is and on what we mean by language (another definition problem). Most species communicate. A dog howling in pain produces a sound which conveys meaning to other members of its species, and other species. Whining, barking and growling all do the same. Is this not a language?

The dog is using sound to convey information about its emotional state. Other species may use language to convey facts. We do this more than any other but we are not alone. Dogs can also learn the meaning of words like 'dinner', 'walkies', 'cats' and 'sit'. These words do not convey emotions, they convey information. It is not just mammals that can communicate in this way. Bees that find a good source of nectar can return to the hive and tell other workers where this is. They convey these facts by the way they move and the way they orientate their bodies as they move. Effectively, they communicate facts by dancing. The workers that have no personal knowledge of the source of the nectar can then fly straight to it. This ability to transfer knowledge is the power of a fact-based language, rather than an emotion-based language.

For most animals knowledge is the result of personal experience (which can include watching other animals and learning by observation). With the development of language, knowledge can result from the experience of other individuals. We no longer need to experience or watch something ourselves to know about it. We know electricity is dangerous and we know it will jump from wires with no insulation – we have been told about this. We do not have to be electrocuted before demonstrating our intelligence in not handling bare live electrical cables.

It may be impossible to say whether having a sophisticated language proves we are intelligent, but having a language greatly increases our ability to look as though we are intelligent because it allows us to gain knowledge out of all proportion to our experience.

Intelligence versus intelligent behaviour

I'm falling into the trap of saying we don't know what intelligence is, but we can look as though we have some. This can't mean anything until we try again to find a definition. If it is not just having a lot of knowledge, then might it be what we do with our knowledge? Could intelligence be the ability to take knowledge and apply it to situations in such a way

YOUR BODY

that it generates behaviour that can adapt to change? Is an intelligent animal an adaptable animal?

We are possibly the most adaptable animal ever to have lived, as we noted in the opening paragraphs of this chapter. To any onlooker this adaptable behaviour certainly suggests intelligence, but so far I have been talking about intelligent behaviour and intelligence as though they are the same thing, and they are not. It is easier to build a robot that *looks* intelligent than it is to build a robot that *is* intelligent. Intelligent behaviour is behaviour that is appropriate to the circumstances – appropriate because it enhances survival or appropriate because it achieves some other desired objective. We can build a robot that steers itself through a maze, yet we know robots are not intelligent (we can't yet build intelligent robots). When an animal steers itself through a maze, we use that as an indication of its intelligence, but what we are really observing is intelligent (or correctly 'apparently intelligent') behaviour, not necessarily intelligence. In this case, we only assume there is intelligent thought behind the apparently intelligent behaviour. Our adaptable behaviour is just intelligent behaviour, not a manifestation of intelligence as such. Interpretation can also be complicated by the fact that behaviour can be viewed from differing perspectives. On the one hand, a species that devises a way of vacuum packing its food in durable tins that keep it edible for years can be viewed as highly intelligent. On the other hand, a species that locks its food into sealed metal containers that can't be opened with any part of its body and are only accessible using a piece of machinery that does not come with the container can look remarkably stupid.

Intelligence as a way of thinking

If intelligence is not a way of behaving, could it be a way of thinking? We can take the facts at our disposal and use them to infer what has happened or what is going to happen, making deductions and predictions about things we did not see and about which we have no information, even second-hand. This is undoubtedly a powerful tool but it has the feel of something a computer could do if it had enough information and was programmed with enough 'IF, THEN' commands (*Deduction*: IF the cake disappeared when there was only one person in the room, THEN that person ate the cake. *Prediction*: IF the sun rose today at about 6 a.m., THEN the sun will rise tomorrow at about 6 a.m.). Given enough information and enough time, it would not seem to require much intelligence to make deductions or predictions like this. However, ideas don't always result from a trudge down the road of deductive logic. Many ideas spring unexpectedly into the thinker's mind as though being injected from elsewhere. This may be a function of our subconscious mind ordering and cross-referencing facts, while the conscious mind considers something else until suddenly the pieces fall into place, but, however it happens, it does not appear to be something a computer is yet capable of. Could this facility be what we call 'intelligence', an ability for our thinking to take shortcuts? Surely this is just what we call having a leap of insight. If we thought it was what we mean when we say 'intelligence', we would not have invented the word 'insight'. If insight is not a reliable indicator of intelligence, is reasoning a better candidate?

Reasoning and intelligence

Reasoning is the ability to solve problems mentally by ordering the information we have and projecting it to a logical conclusion. Reasoning releases us from having to solve problems by trial and error and is another a way of using our brains to short-circuit what would otherwise be a tedious and time-consuming process. For example, if we're looking for grey suits in a clothes store and we know there are blue suits on the second floor, we may reason the grey suits will also be on the second floor. This short-circuits a long-winded attempt to find the grey suits by starting at the entrance door and searching the whole store metre by metre.

YOUR BRAIN

However, reasoning can still look like a trudge down the road of deductive logic. Unlike insight, we are usually conscious of each step and have directed our mind from one to the other until an end is reached.

We tend to feel that people who have an obvious facility for reasoning must be intelligent, but for an observer watching this process the relationship between the reasoning and the degree of intelligence it suggests may not be clear. Although reasoning is the ability to follow a logical chain of thought, there is often more than one possible conclusion that can be drawn from a given set of circumstances and different conclusions may suggest different degrees of intelligence when there is no justification for this.

Here is a simple test of reasoning as may be found in numerous puzzle books and public examinations (although I hope this exact question never appears in any exam).

Complete the next two numbers in the following sequence:
2 4 6 __ __

If three people gave the answers
2 4 6 8 10
2 4 6 10 16
2 4 6 22 116
it is probable that the first person, who spotted they

rise by two each time, would be deemed correct, whereas the other two would almost certainly be deemed to have made a mistake and would be viewed as displaying a poorer ability to reason and probably less intelligence.

However, the second person has merely seen the original sequence as groups of three in which the first two numbers are added to give the third:
given 2 4 6, then 2 + 4 = 6 therefore 4 + 6 = 10 and 6 + 10 = 16

The third person has seen the sequence as pairs in which the two numbers are multiplied, then the difference between them is subtracted from the result to give the next number:
given 2 4 6, then 2 x 4 = 8 but 2 subtracted from 4 = 2 so 8 − 2 = 6

then the next in sequence will be 4 x 6 = 24 but 6 − 4 = 2 so 24 − 2 = 22

the last number must be 6 x 22 = 132 but 22 − 6 = 16 so 132 − 16 = 116

The reasoning of all three is both mathematically correct and logically sound (and if you want something to do on a wet Sunday afternoon there are other solutions that also work). The last two people have not only demonstrated an ability to reason but have also shown an originality not shown by the first. Could this not be seen as an indication of greater intelligence, not lesser? We often respect people who can find new ways of addressing familiar problems.

Whereas insight looks as though it is unrelated to the intellectual power of the thinker, reasoning can be traced from step to step (if we bother to ask the thinker what the steps were) and can appear more impressive because it can be explained. But is 'intelligence' defined by an ability to provide a solution, or an ability to explain how we provide a solution? If the conscious mind solves a problem by reasoning and the sub-conscious mind solves it by insight, neither would seem to have a greater claim to 'intelligence' than the other.

Originality and intelligence

I said we often respect people who can find new ways of addressing familiar problems. We consider an ability to have an idea that no one else has had to be an indication of intelligence, but only if the idea meets two criteria: it must be practicable and it must be socially acceptable.

Someone who suggests dispatching all the world's lawyers to Mars to partition the surface and draw up real-estate deeds ready for later colonisation would not be seen as displaying high intelligence because, no matter how novel the idea (and no matter how socially acceptable), it is not workable. Similarly, Adolf Hitler had original ideas which he proved were practicable but they were socially abhorrent and he is not remembered for his intelligence.

An ability to have a novel, workable, acceptable idea is, however, just seen as an indication of intelligence. It is not intelligence itself.

Imagination and intelligence

Closely related to originality is imagination. We value our imaginations (although we must acknowledge we have no way of assessing the imaginings, if any, of other species). Imagination is a powerful tool. It allows us to build a training ground in our mind and to plan for a variety of events in a safe and harmless environment. We can use our imaginations as a life simulator where the worst can happen but no one gets hurt. We can therefore prepare ourselves for an unknown future and make a safe choice in the real world when we encounter circumstances for the first time. As a consequence, our imagination gives our behaviour the appearance of an underlying intelligence, but (again) it is not intelligence itself.

YOUR BRAIN
Technology and intelligence

Sometimes we use our imagination for considering more than just probable scenarios. Sometimes we imagine the improbable. We imagine being able to breathe underwater, or being able to walk on the moon, then we use this as a springboard for action and make it real through the application of technology.

We may feel our intelligence is most clearly demonstrated by this command of technology but we are not the only species that uses tools. Some chimpanzees will use a blade of grass to draw termites out of a termite mound; other chimpanzees will use two stones as a hammer and anvil to crack tough nuts. European thrushes will place a snail on a stone and use that as an anvil while it breaks open the shell with its beak, and some birds will use a small twig to winkle insects out of holes in the bark of trees.

Two things set our technology apart from that of other species. They use natural objects (carefully selected but not modified) while we modify or create our tools, and they use tools only for necessary activities (usually feeding) while we so delight in our command of technology that we use it for both necessary and unnecessary activities (like smoothing our clothes or making garden gnomes). We are able to do this because for most of us technology has taken over so many of the chores of survival – providing water in our homes and food and clothes in our local shops – that we can now devote significant amounts of our time to non-essential pursuits. Newton and Einstein were freed from the labour of hunting and gathering or building shelters, and were allowed to apply their minds to non-essential matters such as the nature of gravity and what it would be like to travel at the speed of light. With our species' potent combination of knowledge, curiosity and imagination set free by technology, we began to explore all aspects of the universe in an effort to understand everything.

We are the only species that uses technology to achieve our dreams but perhaps what truly makes this a human characteristic is not the fact that we have the technology to do it, but the fact that we have the nerve to do it. Our unquestioning confidence that we can take the wildest leaps of our imaginings and make them real is our greatest strength, and perhaps our greatest peril.

Nevertheless, our technology only demonstrates our intelligence; it does not define it.

Genes and intelligence

Is intelligence defined by our genes? Two hundred years ago in the UK, the working classes were deemed stupid and incapable of being educated (albeit deemed by those who were not working class and who had been educated). At that time, Scotland had four universities, England had two and Wales had none. Today, these same three countries have about 100 universities and dozens of university-sector colleges running degree-level courses. Today, we – or perhaps I mean I – accept that virtually anyone can have a successful university education if they are interested in one and it is made available to them, yet most of today's graduates are descended from the working classes of 200 years ago and have the same genes as people formerly considered unteachable.

There is no doubt everyone's brain is different; variation is the raw material of natural selection. It is as though different brains are wired differently. Some people find languages very easy to learn, others find mathematics or technical drawing easy. We are all different. When we find something we are good at and it coincides with some motivation, we can appear very intelligent. If we have not found a subject we are good at or are not motivated, we can appear intellectually dim.

Many years ago, I worked on the railways with a railman called Bob. During a break on a cold winter nightshift, the conversation drifted to school days. Bob admitted that at school he had hated mathematics; he had been hopeless at arithmetic and had comprehensively failed his exams. Having made his confession he then turned to the horseracing pages of his *Sporting Gazette* and – in his head – commenced a calculation of the odds and potential winnings on a three-race accumulator that would have given a respectable super-computer a migraine. Bob hated maths but this wasn't maths, this was betting and Bob loved betting. It is doubtful Bob's former maths teacher saw him as a star of mental arithmetic but it is also doubtful three-race accumulators loomed large on the maths syllabus. Sometimes people are as clever as they want to be.

So... what is 'intelligence'?

I have no definitive answer to this, and nor has anyone else, but here is an opinion. Intelligence has nothing to do with our behaviour. It is not something that manifests itself outwardly. Intelligence, if it exists at all, is something inside. It is not even a way of thinking, it is the fact that we can think. As I write this, I am struggling to structure my thoughts. I am not reacting to the world around myself; I am looking inwards and wrestling with abstract concepts. We all do this and this ability may be what we think of as intelligence. Intelligence is a facility and we all have it. It cannot be measured and individuals cannot be compared. How it appears to the outside world may differ from person to person but that is a difference in the application of intelligence not in its nature.

Earlier I suggested most of us would intuitively think of intelligence in the same way we think of strength. After exploring the subject, perhaps it would be better to say intelligence is a concept more like 'society'.

Society is a quality that arises in human populations as they become increasingly complex. It does not exist independently of populations and it is not something people decide to create. A society comes into being automatically once more than two or three people congregate and interact, and a society of 20 individuals looks different and behaves differently to a society of 20,000 individuals. The larger a population becomes, the more complex its society becomes. We can compare the qualities of different societies but not the degree of 'society' in each. To put it another way, we cannot compare 'society-ness'.

'Intelligence' may be similar. It is something we recognise as a function of our brain's complexity. We all have it and we can compare what people do with it, but not what it is. No person has more intelligence than another just as no population has more society than another.

We talk about intelligence as though it is inherited and fixed, as though there are people with a high intelligence and there are people who are stupid. If intelligence is just an ability to think – something most animals can apparently do – then it is inherited, but people cannot have a high or low amount of it. I'm going to call what we refer to as high intelligence by a different name, 'cleverness', purely to distinguish it from this definition of intelligence.

There are undoubtedly clever people, but cleverness relies on our having skill and at least some knowledge, and cleverness can be improved with practice. Even Einstein went to university to acquire knowledge and to practise applying it, but no one will acquire knowledge or practise unless they are motivated. Before we categorise people as stupid, we should ask how motivated they are. When we find something we are good at and it coincides with motivation, cleverness can be the result. We are all potentially clever at something. We can be clever because underpinning that possibility is the presence of intelligence – the ability to think.

Most animals, and apparently all vertebrates, have intelligence. Plants and fungi evidently have no intelligence because they have no brains, but brains may not be a necessary prerequisite.

Artificial intelligence

Intelligence evolved as brains became more complicated. The human brain is very complex and this may be the reason why human intelligence also appears very complex. We don't just think, we think to the extent of being self-aware and can pose abstract mind-bending questions such as, 'Do we really exist?' If this is just a function of our brain's complexity, then there should be no bar to a computer being capable of this if it is sufficiently complex – if it crosses a threshold of complicatedness. We are, after all, just biological machines. There is no real difference between a brain made from cells with electrical nerve pathways, and a computer made from metal and crystals with electrical circuits. To compound this, the boundary between brains and computers may soon become blurred. With the advent of the biotechnology revolution, we may eventually be growing biological computers for our offices rather than constructing boxes of plastic and metal and using other vastly inefficient production methods. This should warn us that the self-aware computer is on its way. As usual, science fiction has already explored this possibility, but eventually we in the real world will have to ask, 'Should we build self-aware computers with all the responsibility that that brings?' (But, as we don't know where the threshold of complexity for self-awareness is, they will probably have arrived long before we ask that question.)

Chapter Fifteen

Future Evolution of the Human Body

Natural selection never sleeps. Therefore, its result, evolution, never sleeps. Presumably, our body and its workings will continue to change, but change into what? In the science fiction of the 1950s, it was popular to portray humans of the future with bodies that were a continuation of past trends. Our brains have been increasing in size over the last few million years, so heads of the future were imagined even larger. We have been getting taller, so people of the future were pictured even taller.

YOUR BODY

How we will not look in the future

Sadly predicting the future is not this easy. Evolution has no inner motor pushing it ever further along the same road. For evolution, tomorrow is never an automatic continuation of today. If we are to understand how our bodies will change tomorrow, we must try to anticipate the actions of natural selection in tomorrow's world.

To address this it is possibly helpful to consider the ways in which our ancestors were the subjects of natural selection in yesterday's world and ask if these types of selection are still having a significant effect now.

FUTURE EVOLUTION OF THE HUMAN BODY

Natural attrition

It is not difficult to think of an animal being killed by predators. In our distant African past, our ancestors undoubtedly fell victim to such carnivores, as many people do today in some parts of the world. Hundreds of thousands of years ago, deaths from this cause may have been related to some physical attribute (such as length of legs), and the elimination of some individuals may have had an effect on the physical evolution of the survivors. This was possible when the total world population of hominids or hominid ancestors was small. However, predation will certainly not have a significant effect today when the global human population is around 6,000 million individuals and many predators have been driven almost to extinction. The same can be said of deaths from venomous animal bites and stings, or from eating poisonous plants. We can dispense with accidental deaths at the same time. There are simply too few of any of these, and for all these categories today the victims are unlikely to share any common physical attribute that is being eliminated. Those of us who survive such events are no longer the fittest, only the fortunate.

Social support

We might think our social organisation would act as a damper to future evolution of our bodies, that natural selection would find it harder to remove individuals from a population if those individuals were being supported by their community. However, we share a social habit with our nearest living relatives the chimpanzees and the gorilla and it appears the common ancestor of all of us was already a social animal. This means that our human bodies have become human, the chimpanzees' bodies have become chimpanzee and the gorillas' bodies have become gorilla after we all became social animals. The development of complex interpersonal relationships and the mutual support and benefits to survival that these bring did not halt the changes in our physical appearance as our three groups settled into their different ways of life.

Mate selection

One way our bodies may have continued to change after we became social animals was through the agency of mate selection. This is a common practice among animals. Few species mate with the first member of the opposite sex they encounter during the breeding season. In some species, females will only mate with the dominant male of a group. The dominant male will therefore father all the offspring of the group and the next generation will tend to resemble him in appearance. In such species, males

FUTURE EVOLUTION OF THE HUMAN BODY

will compete with other males to maintain their dominance (stags will lock antlers; lions will fight). There is no reason to believe our primate ancestors lived in this way but other forms of mate selection exist. In some species, females may have a more active say and males may compete with each other in elaborate displays in an attempt to entice desirable females into choosing them. This is common in birds, but in mammals too it is not unusual for the female to be the one selecting her mate. In today's human societies based on the European model, it is the man who traditionally asks a woman if she will marry him. This looks like a man selecting a woman but actually he only decides who to ask, it's the woman who decides what the answer will be. If females in our early history consistently chose males for a particular attribute, that attribute would have tended to become common in the next generation. It may have been mate selection that led to the evolution of the wide range of hair colours displayed by Europeans, the palest of the racial groups. However, whether or not mate selection played a part in our early evolution, it is difficult to see how it could affect our future evolution given the huge population involved and the lack of consistency worldwide in the characters being considered important for partners – especially physical characters.

YOUR BODY
Race

Not only did our bodies continue to change in form after we developed social habits but they also continued to change as some of our ancestors left Africa and colonised the globe. The result was the racial groups we see today, which developed in response to the effective geographical isolation of different populations in different environments. Could these races become even more distinct in the future?

The main argument against this happening is the recent development of technologies such as clothing, housing and agriculture. These separate us from the elements of nature that steered our bodies towards racial distinctiveness. With the growth of international communications, trade and the relatively free movement of ideas and information, we now tend to live under similar conditions wherever we are on the planet. Some of us may be more affluent than others, but this is not significant for the evolution of our bodies.

Obviously, in some restricted areas of the planet, there is an active dilution of racial differences as children are born to inter-racial marriages. This is especially noticeable in the USA and some parts of South America, but, in the context of a global population of 6,000 million (and growing), the number of children born to parents with different racial heritages is tiny.

It is probable human races will not now continue to diverge, although there is no reason to expect racial differences to fade in the foreseeable future.

Culture

With the physical changes came changes in cultural emphasis. In some groups technology advanced quickly; in others, social organisation became increasingly complex. It is wrong to call people who lived, or still live, in mud huts or tents 'primitive'. It is obvious they lack high technology but what is not so apparent is that their traditional social systems may be much richer and more involved than the depauperate and shallow family groupings of, for example, technological Northern Europe. Northern European society has rejected elaborate kinship relationships in favour of the nuclear family. Can the average European name their eight great-grandparents or any of their second cousins?

These different emphases have produced different human societies in different regions of the globe. Is it possible these cultures could remain separate and eventually develop into separate human species with noticeably different bodies?

Historically, humans formed separate groups very

readily. Race, cultural differences and the wide diversity of languages make this clear. Even within language groups, people living only tens of kilometres apart talk quite differently and quite identifiably, even in affluent cultures where transport is readily available and population mobility is high. However, languages and the tendency to have children with partners from the same language group have never led to the splitting of *Homo sapiens* into two or more species, with accompanying changes in their bodies. The differences that led to the appearance of human races over the last 60,000 years are still superficial, and language groups outside Africa must have been distinct for a much shorter time than that. People have simply not been separated into distinct groups long enough to form several species.

Today, the world is increasingly drifting towards a global culture and we can expect any cultural influences on the evolution of our bodies to decline, not increase.

Disease

A much better candidate for future editing of individuals across the whole human species is disease. Natural selection at the hands of disease was probably a hazard for our ancestors and disease is certainly one

FUTURE EVOLUTION OF THE HUMAN BODY

of the major killers worldwide today. The difficulty with making predictions about how disease-related deaths may edit our species is that humans can die from a wide range of conditions and very few of them seem to be related to any anatomical aspects of our bodies. For example, whether you catch malaria does not depend on your shoe size or whether you have a receding chin.

With the onset of global warming and the increase of international air travel the natural spread of diseases and disease-carrying insects into areas not previously affected can be expected to rise, with potentially harmful consequences for cultures and continents with no natural resistance to them. If this causes serious reductions in the childbearing capacity of whole generations, there could be an impact on human physical evolution in some areas. However, it is unlikely this could occur to any noticeable degree.

If natural selection at the hands of disease did bring evolutionary change, it would probably be internal alterations in our immune systems or biochemistry. Even then it would have to be a massive global epidemic of terrifyingly lethal proportions to redesign the survivors of more than 6,000 million people, and this seems very unlikely. At least, we all hope it's unlikely.

Also countering this would be our pharmaceutical technologies which we hope could save the victims, but if these operated as at present – for example during the current AIDS epidemic – the worldwide availability of drug therapies would actually be patchy at best and non-existent at worst. AIDS is now the world's fourth-largest killer after heart disease, strokes and respiratory infections (all of which mainly affect the elderly). By the end of 2005, more than 25 million people had died from AIDS and approximately 40 million were infected with the virus, most in sub-Saharan Africa. The number of new infections is still growing worldwide and it remains to be seen where this epidemic will end, but even this does not seem to be reaching levels where it could change the appearance of our species, although it has changed much behaviour.

War

Like disease, war has the capacity to kill millions. In one of its hideous guises, currently called 'ethnic cleansing', it can involve the mass extermination of targeted groups. However, while this may have devastating consequences for the groups and in theory for their genes, with a world population of 6,000 million, it is unlikely such wars could alter the physical form of the whole species.

FUTURE EVOLUTION OF THE HUMAN BODY
New inventions

The environment today is not the environment in which our bodies originally evolved. We have modified the atmosphere with chemical emissions. We increasingly alter the chemical constituents of our food with pesticide residues, preservatives and artificial flavourings. Each of our bodies is bombarded 24 hours a day with electromagnetic radiation (in other words radio and radar signals and similar emissions *radiating* from a transmitter, not nuclear radiation in the sense of Uranium – don't panic). Our bodies are soaked by radio and television transmissions across a wide range of frequencies from satellites and local masts; from mobile telephone and other communications transmitters; and by electrical fields from the myriad of electrical devices with which we surround ourselves, from electric lighting and vacuum cleaners to kitchen appliances and computers. We are oblivious to this barrage because we have never evolved sense organs that detect it; it did not exist until the 20th century. Will we now evolve senses for this? Only if it affects our reproduction. It could do this by affecting our health early in life or by affecting our reproductive systems directly. There is no evidence at present that it does either but it would be folly not to continue to monitor these new environmental factors and their effects.

YOUR BODY

Electricity has already altered our lives in ways our evolution never anticipated. By using cheap and readily available bright artificial light, we can now extend the length of the day regardless of the time of year, but this does have an effect. Our bodies naturally react to light levels. At temperate latitudes in the winter months, when day length is severely shortened, a person without access to artificial light will often sleep for up to 16 hours. With artificial light, this can be more than halved. The use of electric light to extend the day artificially can lead to disruption of the normal pattern of hormone production. Hormones are the body's chemical messengers, altering our overall physiological state or affecting specific organs, and some hormones have a daily rhythm of high and low production. Even if this disruption does not happen as a direct response to the light, it may still occur as we maintain our activity beyond what would otherwise be normal and override our body's internal clock. We also disrupt this clock when we make long-distance air journeys which disturb our normal day/night cycle.

We feel the effect of these disruptions because our body, including our metabolism, has evolved in a stable cycle of day and night, light and dark. Whether our ability to disrupt the body's natural rhythms will have an evolutionary effect depends on whether enough people are affected and whether the disruption interferes with our ability to have children. Even if thousands of millions of people were affected and it did alter their reproductive capabilities – neither of which is at all probable – there would not appear to be any common genetic heritage shared by people who habitually work late at the office or travel by jet so we could expect no effect on the future evolution of our species' anatomy. It's not as though only people whose eyes are very close together habitually work late, or only people with no earlobes who take international flights.

Genetically modified organisms

As a species, we have recently developed the ability to alter the genetic composition of other species directly, to produce Genetically Modified Organisms (GMOs). Some people argue this is no different to altering the characters of other species by selective breeding, which we have been doing for thousands of years, but this is nonsense. Selective breeding is a way of choosing which cow genes we want to see in our cows, or picking particular dog characteristics for our dogs. With genetic modification, we are moving genes from one species to another. When we take the gene from a fish that produces a natural anti-freeze in its blood and put it into a tomato to stop the tomato freezing during refrigerated transport, we are doing something that would be very difficult to achieve by selective breeding.

There is much dispute worldwide about the wisdom of such activities. Genes are communities of molecules that have evolved together. They act in concert and the tissues they create communicate with each other to produce the final result. It requires a great deal of very stringent and safety-conscious research to ensure any inserted gene will produce only one simple outcome that does not affect the rest of the host organism, or future generations of that organism.

Many people express concern about the possible impacts of GMOs on human health or on the environment. Eating modified DNA from GMOs or products made from them is unlikely to affect our health; humans have always eaten DNA from other species. Every time we eat a banana or a chicken, we eat foreign DNA; we simply digest it. We have evolved to digest DNA and artificially created gene combinations can expect the same fate. However, eating the modified products themselves may be a different matter. Modifying the genetic composition could in theory produce resulting modifications in the cells, generating chemicals not naturally occurring in the unmodified plants. These might affect long-term health or produce allergic reactions.

Presumably, laboratories creating GMOs test their products for both these potential effects in clinical tests before initiating field trials of GMO crops.

If human-made DNA escaped from genetically modified crops into wild plant populations, the consequences for the environment would be impossible to predict. Nothing similar to this has ever been possible in the past — until now, laboratory-modified DNA has never existed. Consequently, no one has any relevant experience on which to base a risk assessment.

Within the context of this book, the question must be whether the escape of new gene combinations into the environment at large could affect the course of human physical evolution. This is impossible to gauge. Any damaging effects might be restricted to the environment — where they would almost certainly be irreversible — but how our species might be affected by any such environmental change cannot be anticipated. Again, no one has any experience on which to base a prediction.

Genetically modified humans

Doctors currently use many medical techniques to help couples have children. Fertility experts can use drugs to enhance ovulation, in vitro fertilisation (IVF treatment) and storage of frozen embryos with later implantation. Couples who could not have passed their genes to the next generation without this assistance can now have healthy babies. It can be argued that medical assistance in the reproductive process is not unnatural. Our species is a product of nature; nothing we do can ever be unnatural. Even burying the countryside under a sprawling city of concrete and glass is no more unnatural than other species burying the seabed under a sprawling coral reef, or beavers flooding the landscape with a dam. Some people consider medical intervention in the reproductive process undesirable, but desirability is a separate issue. Doctors only exist to make the body work when it isn't working by itself. We seldom ask doctors not to intervene in the natural bleeding process or the 'dying from influenza' process.

Nevertheless, we now stand on the verge of a very different form of intervention: genetic engineering and conscious selection of characters for our children. We have always taken decisions which to some extent affect the appearance of our children but we have done this through the practice of mate selection ('~~Must be bald~~') or in some cases the more recent capability to select the characteristics of otherwise anonymous sperm donors. We can now anticipate being able to select the features of our children directly with the advent of so-called 'designer babies'.

FUTURE EVOLUTION OF THE HUMAN BODY

It could be said that choosing what we want the next generation to look like is not unnatural ('nothing we do can ever be unnatural') but there are inherent dangers for designer babies. Choosing characters today assumes we know what will be important tomorrow. We may inadvertently choose to eliminate characters which will later prove critical for survival.

Designer babies have been proposed before but, in the absence of genetic engineering, it was within the context of human selective breeding. This manifested itself in the eugenics movement of 1930s Europe and was most forcefully expressed in the push by the Nazis to create an Aryan super-race with blonde hair and blue eyes. As well as its truly horrifying rejection of human variation, this was a fundamentally flawed concept as it assumed the environment would never change. Seventy years later, as the holes in the ozone layer grow, blondes will be more susceptible to skin cancer, and blue eyes function less well than brown eyes in high light intensity. If we keep damaging the atmosphere, it may be the dark-skinned, dark-haired, dark-eyed individuals who maintain their health and produce more children, while the numbers of blue-eyed blondes in the world falls from generation to generation as health problems early in life interfere with childbearing. This, of course, is unlikely to happen and not just because blue-eyed blondes tend to live in those parts of the world where people can afford sunblock, sunglasses and high-quality healthcare, but the fact remains that the Nazis did not anticipate a hole in the ozone layer.

Similarly, we cannot anticipate tomorrow's environment as we prepare to design babies to face it, but without that knowledge what criteria are we using to make our choices? When we talk about choosing the eye colour or the height of our child-to-be, are we really suggesting what would be best for the child or are we choosing what we would have preferred to look like in a society obsessed by physical stereotypes? Surely selecting a more efficient immune system for a child should warrant a higher priority?

Although designer babies may become a reality and we may be able to influence the shape and attributes of the human body as a result, if this form of direct genetic manipulation does pass the test of social acceptability, it can be expected to do so only for some of the people in some of the wealthier nations. It is unlikely to have the potential in the foreseeable future to affect the evolution of the human body globally but it could allow the development of localised genetic caste systems, a theme already explored by science-fiction writers.

FUTURE EVOLUTION OF THE HUMAN BODY
Planetary birdstrike

In trying to predict how the human body may change in the future, we keep coming back to the same problem. The situation today with such a numerous population spread across the entire planet is completely different to the circumstances that existed when *Homo sapiens* first evolved from a small number of individuals on the African continent. Anything that is to affect the physical appearance of our species would have to act across vast distances and involve immense numbers of people. At present we cannot even begin to speculate what that might be. It may take nothing less than a global catastrophe on the scale of a collision with an asteroid or comet, or a massive solar flare, wiping out most humans and leaving only small survivor populations to repopulate the planet. Only this may now have the capacity to alter the appearance of every member of our species. If we are fortunate, therefore, our species may not change noticeably in appearance in the future, although as evolution never sleeps this may yet be a naive view.

We should not be surprised by this inability to predict what will happen to our bodies. We are part of life, and there is no greater complexity known than life. The physical sciences (physics, chemistry, geology, astronomy – even meteorology) deal with the simple end of nature; it is biology that deals with the ultimate complexities, complexities often beyond the capabilities of physical laws and mathematics to predict. Biology has too many variables for there to be certainty. If I drop a rock, it will fall to the ground. If I drop a bird, who knows where it will land?

Evolution of the non-human body

Let us conclude this chapter of speculation by taking a trip off-world. The diversity of life on Earth is vast but other planets and their moons have very different environments. The Earth is unique even within our own solar system. When biologists try to predict what life might be like on other planets, they invariably mean life like Earth's, but no one can know how life may manifest itself across the galaxy. In other star systems, there may be crystalline life forms with life spans of 10,000 years, or molecular life forms with life spans of fractions of a second. Our predictions are shackled not just by our ignorance of the universe but also by our familiarity with the Earth. We find it hard to think beyond our own experience.

If this book teaches us anything, it teaches us that our body is the result of a long and complex history during which numerous changes of direction occurred, any one of which could easily have been different. We only look the way we do because we arrived by way of: bilateral symmetry; jaws and teeth;

FUTURE EVOLUTION OF THE HUMAN BODY

paired fins; four limbs; five digits; two eyes; elbows and knees; wrists and ankles; life in trees; four paws as hands; loss of tail; walking upright; and rear hands becoming feet again.

This in turn tells us that whatever extra-terrestrial beings look like they will not be little green men or almond-eyed 'greys'. The odds of a different evolutionary history producing a humanoid body are astronomical. We only have to look around at other animal species on Earth to see how easily evolution produces different body shapes. Just compare a person with an octopus or an insect or an earthworm or a jellyfish. No, if a flying saucer lands and little green men emerge, we can be sure of two things – they are from Earth and their ancestors were fish.

How aliens will not look